Mathematics Framework for California Public Schools

Kindergarten Through Grade Twelve

Developed by the
Curriculum Development and Supplemental Materials Commission

Adopted by the
California State Board of Education

Published by the
California Department of Education

Publishing Information

When the *Mathematics Framework for California Public Schools, Kindergarten Through Grade Twelve* was adopted by the California State Board of Education on December 10, 1998, the members of the State Board were the following: Yvonne W. Larsen, President; Robert L. Trigg, Vice President; Marion Bergeson; Timothy Draper; Kathryn Dronenburg; Marion Joseph; Marion McDowell; Janet G. Nicholas; Gerti B. Thomas; Marina Tse; and Richard Weston.

The framework was developed by the Curriculum Development and Supplemental Materials Commission. (See pages viii–x for the names of the members of the Commission and the names of the principal writers and others who made significant contributions to the framework.)

This publication was edited by Janet Lundin and Sheila Bruton, working in cooperation with Catherine Barkett, Administrator, and Mary Sprague, Consultant, Curriculum Frameworks and Instructional Resources Office, California Department of Education. It was prepared for printing by the staff of CDE Press. The cover and interior design were created and prepared by Cheryl McDonald. Typesetting was done by Carey Johnson. The framework was published by the California Department of Education, 721 Capitol Mall, Sacramento, California (mailing address: P.O. Box 944272, Sacramento, CA 94244-2720). It was distributed under the provisions of the Library Distribution Act and *Government Code* Section 11096.

© 1999 by the California Department of Education
All rights reserved

ISBN 0-8011-1474-8

Ordering Information

Copies of this publication are available for $17.50 each, plus shipping and handling charges. California residents are charged sales tax. Orders may be sent to CDE Press, Sales Office, California Department of Education, P.O. Box 271, Sacramento, CA 95812-0271; FAX (916) 323-0823. Current information on prices, credit card purchases, and shipping and handling charges may be obtained by calling the Sales Office at (800) 995-4099.

An illustrated *Educational Resources Catalog* describing publications, videos, and other instructional media available from the Department can be obtained without charge by writing to the address given above or by calling the Sales Office at (916) 445-1260.

Prepared for publication
by CSEA members

Contents

Foreword iv
Preface vi
Acknowledgments viii
Introduction 1

1 **Guiding Principles and Key Components of an Effective Mathematics Program** 5

2 **The California Mathematics Content Standards** 17

  Kindergarten 22
  Grade One 26
  Grade Two 30
  Grade Three 35
  Grade Four 43
  Grade Five 52
  Grade Six 57
  Grade Seven 64
  Introduction to Grades Eight Through Twelve 72
  Algebra I 74
  Geometry 78
  Algebra II 83
  Trigonometry 87
  Mathematical Analysis 90
  Linear Algebra 92
  Probability and Statistics 94
  Advanced Placement Probability and Statistics 96
  Calculus 99

3 **Grade-Level Considerations** 104

  Preface to Kindergarten Through Grade Seven 109
  Kindergarten 112
  Grade One 116
  Grade Two 120
  Grade Three 126
  Grade Four 131
  Grade Five 137
  Grade Six 142
  Grade Seven 148

  Preface to Grades Eight Through Twelve 153
  Algebra I 158
  Geometry 162
  Algebra II 168
  Trigonometry 172
  Mathematical Analysis 174
  Probability and Statistics 176

4 **Instructional Strategies** 178

5 **Assessment** 194

6 **Universal Access** 201

7 **Responsibilities of Teachers, Students, Parents, Administrators** 210

8 **Professional Development** 216

9 **The Use of Technology** 222

10 **Criteria for Evaluating Instructional Resources** 228

Appendixes

  A. Sample Instructional Profile 237
  B. Elementary School Sample Lesson Plan: An East Asian Approach 245
  C. Middle School Sample Lesson Plan: Compound Interest 254
  D. Resource for Secondary School Teachers: Circumcenter, Orthocenter, and Centroid 279
  E. Sample Mathematics Problems 297

Glossary 330

Works Cited 335

Additional References 339

Resources for Advanced Learners 342

Foreword

Mathematics—using abstract symbols to describe, order, explain, and predict—has become essential to human existence. A vast body of mathematical knowledge and understanding has been developed over the years and has been infused into every field of human endeavor—science, humanities, arts, commerce, agriculture, transportation, sports, even religion. Mathematics is as important as oral and written language in the communication of ideas.

Mathematics enables us to conduct the simple yet vital transactions of daily life, such as telling time, gauging distances, and making change. It also makes possible complex operations beyond the imagination of earlier generations, such as landing sophisticated exploration equipment on distant planets. At both extremes and in the full range between, mathematics presents a singular beauty. It appeals uniquely to the human mind. Teachers, in particular, understand this beauty and appeal when they see students' eyes alight with recognition and comprehension. Every student can achieve mathematical competence and literacy, though not all will choose to pursue advanced mathematics study, such as topology or number theory. Every student is entitled to a solid foundation in mathematics that includes algebra, geometry, and probability and statistics.

Over the past two decades, there has been considerable experimentation and theorizing about the teaching of mathematics. Differing but sound pedagogical alternatives have been identified. Unfortunately, insufficient attention has been devoted to content—what teachers need to teach and what students need to learn—in kindergarten and each of grades one through seven so that students may be fully prepared for higher-level mathematics beginning in grade eight. Vagueness in content has plagued both instruction and assessment.

This framework focuses squarely on content, as defined by the grade-level and discipline-specific content standards adopted by the California State Board of Education in December 1997. It calls for instructional programs and strategies, instructional materials, professional development, and assessments that are aligned with the standards. Because the standards outline a more ambitious and challenging content than has previously been expected of students, this framework provides instructional guidance to teachers to enable them to raise their benchmarks for achievement and mastery in realistic ways. The framework includes sample problems to illustrate the standards. It also discusses common misconceptions and pitfalls regarding standards-based education and ways to ensure that students are ready to move from grade to grade and discipline to discipline.

The framework provides research-based information about how children learn and general guidelines about instructional strategies. Relatively new teachers, in particular, will find this document useful in providing a solid foundation for mathematics instruction that fits the material being taught. Veteran teachers will find it a source of refreshing ideas to help apply their experience in new ways that will enhance the mathematics achievement of their students. The framework stresses careful and thoughtful sequencing of instruction based on the standards so that every student in every school will master basic skills and techniques for sophisticated problem solving and understand sound mathematical reasoning.

The role of mathematics in human existence—personal, social, and economic—will become even more important in the coming millennium. California students deserve no less than to be educated in mathematics to levels consistent with their counterparts throughout the world. This framework maps the route toward that very achievable goal as set forth in the *Mathematics Content Standards for California Public Schools*. It is a journey that we have no choice but to commence and no alternative but to complete successfully. For our children's future, please turn the page . . . and take the first step.

DELAINE EASTIN
State Superintendent of Public Instruction

YVONNE W. LARSEN
President, California State Board of Education

Preface

> This framework is based on the rigorous mathematics standards adopted by the California State Board of Education in December 1997.

California mathematics educators, parents, and policy makers unanimously agree on the ultimate goal for California students in mathematics: All students need to develop proficiency in mathematics. *Proficiency in mathematics* means that students can solve meaningful, challenging problems; demonstrate both a depth and breadth of mathematical understanding; and perform both simple and complex computations and mathematical procedures quickly and accurately, with and without the aid of computational tools. Many mathematicians, educators, and experts on children's mathematical thinking and learning view these components as ideally interdependent. This framework and the *Mathematics Content Standards for California Public Schools,* on which it is based, focus on realizing that ideal for California students.[1]

The importance of a highly numerate workforce to the continuing growth of state and national economies is rapidly increasing (Paglin and Rufolo 1990). The need for mathematically competent citizens and employees translates into significant employment opportunities for mathematically proficient graduates from California public schools (Grogger and Eide 1995). The corollary is that graduates without mathematical competence will have limited educational and employment choices. One prominent economist estimated that the poor literacy and mathematical competencies of American students will cost the national economy nearly $170 billion per year by the year 2000 (Bishop 1989).

Alignment of the Framework with the Mathematics Standards

This framework is based on the rigorous mathematics standards adopted by the California State Board of Education in December 1997. As required by the California *Education Code,* the framework is aligned with the standards, and it provides a context for a coordinated effort to enable all California students to achieve rigorous, high levels of mathematics proficiency.

The following objectives in this framework replace those in the 1992 *Mathematics Framework*:

- To provide teachers with a clear map of the strategies that are most likely to result in high levels of proficiency for all students
- To give specific guidelines to developers of mathematics instructional resources to ensure that resources are of high quality and are aligned with the mathematics standards
- To foster a balance between the basic computational and procedural skills, problem solving, and conceptual understanding
- To help teachers ensure that students with special needs have access to the resources that will enable them to develop mathematical proficiency

---

[1] The California mathematics content standards, which were adopted by the California State Board of Education in December 1997, are reproduced in their entirety in Chapter 2, "The California Mathematics Content Standards."

Creating an environment in which all students have the opportunity to become proficient in mathematics will not be easy to achieve. To strive for less, however, would shortchange California students and threaten the well-being of our society.

The Process of Developing the Framework

In 1995 Assembly Bill 265 established the Academic Content and Performance Standards Commission (Standards Commission) to develop content and performance standards in the core curricular subject areas, including mathematics. As required by follow-up legislation (Senate Bill 430, Chapter 69, statues of 1996), the content of the *Mathematics Framework* then needed to be brought into alignment with that of the mathematics standards.

In January 1997 the Mathematics Curriculum Framework and Criteria Committee (Framework Committee), appointed by the California State Board of Education, met for the first time to begin the lengthy public process of drafting a new *Mathematics Framework* that would be aligned with the mathematics standards. Simultaneously, the standards were being developed by the Standards Commission. The members submitted their draft of the *Mathematics Content Standards* to the State Board for approval in the fall of 1997, and in December 1997 the State Board adopted statewide mathematics content standards for California. Also in the fall of 1997, the members of the Framework Committee submitted their draft of the *Mathematics Framework* to the Curriculum Development and Supplemental Materials Commission (Curriculum Commission).

The Curriculum Commission distributed more than 3,000 copies of the draft of the framework throughout California's education and mathematics community to gather recommendations from the public on how to improve the draft. The Curriculum Commission carefully considered all the suggestions gathered during the field review and at public hearing sessions. While the draft was being prepared, the Commission worked closely with a number of university mathematics professors, including Hung Hsi Wu, University of California, Berkeley; Ralph Cohen, Stanford University; and Scott Farrand, California State University, Sacramento, who were asked by the State Board to develop problems illustrative of the standards, and with a cognitive psychologist, David Geary, of the University of Missouri. In the fall of 1998, the Curriculum Commission submitted the final draft of the *Mathematics Framework* to the State Board of Education for further public hearing and information. In December 1998 the State Board approved the framework.

LESLIE FAUSSET
Chief Deputy Superintendent
Educational Policy, Curriculum, and
   Department Management

SONIA HERNANDEZ
Deputy Superintendent
Curriculum and Instructional Leadership Branch

WENDY HARRIS
Assistant Superintendent and Director
Elementary Division

CATHERINE BARKETT
Administrator
Curriculum Frameworks and Instructional
   Resources Office

Acknowledgments

The development of the *Mathematics Framework* involved many individuals. Participating directly as members of the **Curriculum Development and Supplemental Materials Commission** were the following members of the **Mathematics Subject Matter Committee (SMC):**

Barbara Smith, Math SMC Chair, Superintendent, San Rafael City Schools

Sheri Willebrand, Math SMC Vice Chair, Teacher/Math Specialist, Ventura Unified School District

Marilyn Astore, Assistant Superintendent, Instructional Support Services, Sacramento County Office of Education

Michele Garside, Superintendent, Laguna Salada Union School District

Viken "Vik" Hovsepian, Teacher/Chair, Mathematics Department, Glendale High School, Glendale Unified School District

Susan Stickel, Assistant Superintendent, Instructional Support, Elk Grove Unified School District

Richard Schwartz, Teacher, Torrance Unified School District

Other members of the Curriculum Commission:

Kirk Ankeney, Chair, Curriculum Commission, Vice Principal, San Diego Unified School District

Roy Anthony, Teacher, Grossmont Union High School District

Patrice Abarca, Teacher, Los Angeles Unified School District

Eleanor Brown, Assistant Superintendent, San Juan Unified School District

Ken Dotson, Teacher, Turlock Joint Elementary School District

Senator Leroy F. Greene

Lisa Jeffery, Vice Principal, Los Angeles Unified School District

Joseph Nation, Managing Director, California Data & Analysis, San Rafael

Assembly Member Jack Scott

Lillian Vega Castaneda, Associate Professor, California State University, San Marcos

Jean Williams, Retired Area Assistant Administrator, Fresno Unified School District

The *Mathematics Framework* reflects the work of a number of **university professors**. The following university mathematics professors developed sample problems to illustrate the mathematics standards included in the framework:

Dr. Ralph Cohen, Stanford University

Dr. Scott Farrand, California State University, Sacramento

Dr. H-H Wu, University of California, Berkeley

Dr. David Geary, Cognitive Psychologist from the University of Missouri, prepared information about how students learn mathematics, the models for mathematics instruction, and the key standards at each grade level for the framework.

Dr. Shin-ying Lee, Center for Human Growth and Development, University of Michigan, provided Appendix B, "Elementary School Sample Lesson: An East Asian Approach."

In addition, **Dr. H-H Wu**, University of California, Berkeley, and **Dr. James Milgram**, Stanford University, performed a technical review of the framework to ensure mathematical accuracy throughout the document.

Initial drafts of the *Mathematics Framework* were developed by the **Mathematics Curriculum Framework and Criteria Committee (CFCC)**. Following is a partial list of the committee membership:

Henry Alder, University of California, Davis

Ruth Uy Asmundson, School Board Trustee, Davis Joint Unified School District

Eduardo Chamorro, Los Banos Unified School District

Paul Clopton, Veterans Affairs Medical Center, San Diego

Ralph Cohen, Stanford University

Mike Erickson, Santa Cruz City Schools

Deborah Tepper Haimo, University of California, San Diego

Viken "Vik" Hovsepian, Glendale Unified School District

Grace Hutchings, Los Angeles Unified School District

Alla Korinevskaya, Temple Emanuel Day School

Patricia Montgomery, Oceanside Unified School District

Calvin Moore, University of California, Berkeley

Martha Schwartz, California State University, Dominguez Hills

Dennis Stanton, Santa Cruz City Schools

Bob Stein, California State University, San Bernardino

Paraskevi Steinberg, Sacramento City Unified School District

Ze'ev Wurman, Dyna Logic, Santa Clara

Subsequent to the development of the initial drafts of the mathematics framework, the State Board of Education adopted content standards in mathematics (December 1997). The commission then revised the draft framework to incorporate the mathematics standards.

The principal writer for the framework, working both with the CFCC committee and with the Curriculum Commission, was **Hank Resnik**.

California Department of Education staff who contributed to developing the framework are as follows:

Leslie Fausset, Chief Deputy Superintendent for Educational Policy, Curriculum, and Department Management

Sonia Hernandez, Executive Secretary to the Curriculum Commission and Deputy Superintendent, Curriculum and Instructional Leadership Branch

Wendy Harris, Assistant Superintendent and Director, Elementary Education Division

Catherine Barkett, Administrator, Curriculum Frameworks and Instructional Resources (CFIR) Office

Tom Adams, Consultant, CFIR Office

Tom Lester, Consultant, Model Programs and Networks

Joanne Knieriem, Support Staff, CFIR Office

Christine Rodrigues, Consultant, CFIR Office

Mary Sprague, Consultant, CFIR Office

Tracy Yee, Staff Services Analyst, CFIR Office

At the time of the adoption of the framework, the following California State Board of Education members supported the Curriculum Commission and provided leadership in the development of this framework:

Yvonne W. Larsen, President, State Board of Education

Robert L. Trigg, Vice President

Marion Bergeson

Timothy Draper
Kathyrn Dronenburg
Marion Joseph
Marion McDowell
Janet G. Nicholas
Gerti B. Thomas
Marina Tse
Richard Weston

A special thanks is extended to **Robert L. Trigg** and **Janet G. Nicholas**, liaisons for mathematics from the State Board of Education.

Appreciation is also extended to:

Susan Greene, Math Specialist, San Juan Unified School District

Nancy Maiello, Teacher, Mira Loma High School, San Juan Unified School District

Introduction

All students need a high-quality mathematics program designed to prepare them to choose from among a full range of career paths. California's Mathematics Task Force (1995) called for a rigorous and challenging mathematics program for every student—a complete program that reflects a balance of computational and procedural basic skills, conceptual understanding, and problem solving. Yet the current mathematics achievement of many students is unacceptably low (Reese et al. 1997). Educators are challenged to expect more from students in mathematics, to raise the bar for student achievement through more rigorous curriculum and instruction, and to provide the support necessary for all students to meet or exceed California's mathematics standards.

To compete successfully in the worldwide economy and to participate fully as informed citzens, today's students must have a high degree of comprehension in mathematics. The new goal is that all students will attain California's mathematics standards for kindergarten through grade twelve (hereinafter referred to as

Introduction

Proficiency in mathematics is a consequence of sustained student effort and effective teacher instruction.

"the standards") and that many will be inspired to pursue advanced studies in mathematics.

This framework is predicated on the belief that proficiency in mathematics is a consequence of sustained student effort and effective teacher instruction. All students are capable of understanding mathematics, given the opportunities and encouragement to do so.

What's New in the 1998 Framework?

The most important feature of this framework is its focus on the content of the mathematics standards adopted by the California State Board of Education. The goal of accelerating student progress through a standards-based program has a significant impact on the entire curriculum. As specified by the standards, much of the content of the mathematics curriculum has been shifted into earlier grades, and mathematics instruction in kindergarten through grade seven is substantially strengthened. These changes necessitate a more substantive, rigorous, and demanding curriculum and more systematic instruction that will better serve California students.

This framework is designed to prepare all children to study algebra at least by the eighth grade, as prescribed by the mathematics standards. The framework describes an introductory algebra course for the eighth grade, followed by the study of geometry in the ninth grade and advanced algebra in the tenth grade, or the alternative of a comparable three-year integrated program. In addition, in accordance with the standards, five strands are now used to describe the curriculum instead of the eight strands used in the 1992 edition of the *Mathematics Framework*.

An important theme stressed throughout this framework is the need for a balance in emphasis on computational and procedural skills, conceptual understanding, and problem solving. This balance is defined by the standards and is illustrated by problems that focus on these components individually and in combination. All three components are essential.

A General Overview: Purpose, Organization, and Audience

The purpose of this framework is to guide the curriculum development and instruction that teachers provide in their efforts to ensure that all students meet or exceed the mathematics standards. The framework provides a context for implementing the standards in the form of guidelines for the design of curricula, instructional materials, instructional practices, and staff development. Building on the standards, the framework addresses the manner in which all students in California public schools can best meet the standards. *All students* includes those performing at, below, and above grade level; English language learners; special education students; and others with special learning needs.

More specifically, the framework:

- Describes *guiding principles and key components* of an effective mathematics program (Chapter 1)
- Presents the essential skills and knowledge expected of students in mathematics as described in the *Mathematics Content Standards for California Public Schools* and illustrated by sample problems (Chapter 2)
- Describes *special considerations and emphases for each grade level to ensure student success* consistent with the

mathematics standards and in anticipation of the statewide testing program (Chapter 3)
- Provides guidance, based on current research, regarding *instructional strategies* and sample lessons that can be effective in ensuring that every child will meet or exceed grade-level standards in mathematics (Chapter 4)
- Guides the development of appropriate *assessment* methods (Chapter 5)
- Suggests specific strategies to ensure *access* to appropriately challenging curriculum for special needs students (Chapter 6)
- Describes the *responsibilities* that all stakeholders must uphold for effective implementation of a rigorous and coherent kindergarten through grade twelve mathematics curriculum (Chapter 7)
- Suggests guidelines for both preservice and in-service *professional development* (Chapter 8)
- Provides guidance on the use of *technology* in mathematics instruction (Chapter 9)
- Specifies requirements for *instructional resources,* including print and electronic learning resources (Chapter 10)

In short, the framework should be viewed as a critical tool for designing and implementing an effective mathematics program in kindergarten through grade twelve and for evaluating instructional resources.

The framework addresses two primary audiences: (1) educators; and (2) developers of instructional resources. Educators include those involved in the day-to-day implementation of school mathematics programs—classroom teachers, school administrators, district personnel, school board members, and others responsible for curriculum and instruction. It also addresses other important audiences, such as parents and community members, including business and civic leaders, who have a vital stake in the success of California students in mathematics.

The following themes permeate the *Mathematics Framework:*

The framework:

➡ **Builds on the mathematics standards** and aligns them with curriculum, instruction, resources for instruction, and assessment, resulting in a coherent and pragmatic plan for achieving high levels of mathematics proficiency for all students. It also provides guidance for understanding the standards by identifying priorities within the standards and offering concrete examples of mathematical problems that demonstrate the concepts of the standards.

The framework*:*

➡ **Emphasizes the importance of a balanced mathematics curriculum.** In particular, the framework stresses the critical interrelationships among computational and procedural proficiency, problem-solving ability, and conceptual understanding of all aspects of mathematics, from the simplest calculations to the most sophisticated problem solving.

The framework*:*

➡ **Addresses the needs of all learners,** with no learner left out and no learner taught at the expense of another; emphasizes prevention over remediation, while recognizing the appropriateness of remediation when it is required; and provides suggestions for instructional

Introduction

The framework is a critical tool for designing and implementing an effective mathematics program in kindergarten through grade twelve and for evaluating instructional resources.

strategies that may be used with students who are English language learners, advanced learners, special education pupils, or at risk of failing mathematics.

The framework:

➡ **Highlights the importance of mathematical reasoning.** The mathematical reasoning standards are different from the others in that they do not represent a specific content area. Mathematical reasoning cuts across all strands. It characterizes the thinking skills that students can carry from mathematics into other disciplines. Constructing valid arguments and criticizing invalid ones are inherent in doing mathematics.

The framework:

➡ **Stresses the importance of frequently assessing student progress toward achieving the standards.** Students cannot afford to wait for a year-end test; rather, they should be assessed frequently throughout the year to determine their progress toward achieving the standards. Teachers, students, and parents need some daily or weekly indication of the extent to which the standards are likely to be met.

The framework:

➡ **Avoids oversimplified guidance on either content or pedagogy** in favor of guidelines on effective instruction derived from reliable research.

1

Guiding Principles and Key Components of an Effective Mathematics Program

A long-standing content issue in mathematics concerns the balance between theoretical and applied approaches. Mathematics is both. In the theoretical (pure) sense, mathematics is a subject in its own right with distinct methods and content to be studied. But mathematics is also extremely applicable both in the practical sense and in connection to other realms of study, including the arts, humanities, social sciences, and the sciences. Any comprehensive representation of the content of mathematics must balance these aspects of beauty and power.

—A. Holz, *Walking the Tightrope*

The major goals of mathematics education can be divided into two categories: goals for teachers and goals for students.

Goals for Teachers

Goals for teachers to achieve are as follows:

1. Increase teachers' knowledge of mathematics content through professional development focusing on standards-based mathematics.
2. Provide an instructional program that preserves the balance of computational and procedural skills, conceptual understanding, and problem solving.
3. Assess student progress frequently toward the achievement of the mathematics standards and adjust instruction accordingly.
4. Provide the learning in each instructional year that lays the necessary groundwork for success in subsequent grades or subsequent mathematics courses.
5. Create and maintain a classroom environment that fosters a genuine understanding and confidence in all students that through hard work and sustained effort, they can achieve or exceed the mathematics standards.
6. Offer all students a challenging learning experience that will help to maximize their individual achievement and provide meaningful opportunities for students to exceed the standards.
7. Offer alternative instructional suggestions and strategies that address the specific needs of California's diverse student population.
8. Identify the most successful and efficient approaches within a particular classroom so that learning is maximized.

Goals for Students

Goals for students to achieve are as follows:

1. Develop fluency in basic computational and procedural skills, an understanding of mathematical concepts, and the ability to use mathematical reasoning to solve mathematical problems, including recognizing and solving routine problems readily and finding ways to reach a solution or goal when no routine path is apparent.
2. Communicate precisely about quantities, logical relationships, and unknown values through the use of signs, symbols, models, graphs, and mathematical terms.
3. Develop logical thinking in order to analyze evidence and build arguments to support or refute hypotheses.
4. Make connections among mathematical ideas and between mathematics and other disciplines.
5. Apply mathematics to everyday life and develop an interest in pursuing advanced studies in mathematics and in a wide array of mathematically related career choices.
6. Develop an appreciation for the beauty and power of mathematics.

It is well known that California students lag behind students in other states and nations in their mastery of mathematics (Reese et al. 1997; Beaton et al. 1996). Comparing the 1990s to the 1970s, a study found that the number of students earning bachelor's and master's degrees in mathematics has decreased during the last 20 years (NCES 1997). At the same time the number of students entering California State University and needing remediation in mathematics has been increasing (California State University 1998). The result of students achieving the goals of this framework and mastering the California mathematics standards will be not only an increase in student mastery of mathematics but also a greater number of students who have the potential and interest to pursue advanced academic learning in mathematics. Because many jobs directly and indirectly require facility with different aspects of applied mathematics (Rivera-Batiz 1992), achieving the goals of this framework will also enable California students to pursue the broadest possible range of career choices.

By meeting the goals of standards-based mathematics, students will achieve greater proficiency in the practical uses of mathematics in everyday life, such as balancing a checkbook, purchasing a car, and understanding the daily news. This process will help the citizens of California understand their world and be productive members of society.

When students delve deeply into mathematics, they gain not only conceptual understanding of mathematical principles but also knowledge of and experience with pure reasoning. One of the most important goals of mathematics is to teach students logical reasoning. Mathematical reasoning and conceptual understanding are not separate from content; they are intrinsic to the mathematical discipline that students master at the more advanced levels.

Students who understand the aesthetics and beauty of mathematics will have a deep understanding of how mathematics enriches their lives. When students experience the satisfaction of mastering a challenging area of human thought, they feel better about themselves (Nicholls 1984). Students who can see the interdependence of mathematics and music, art, architecture, science, philosophy, and other disciplines will become lifelong students of mathematics regardless of the career they pursue.

When students master or exceed the goals of standards-based mathematics instruction, the benefits to both the individual and to society are enormous. Yet achieving these goals is no simple task. Hard work lies ahead. This framework was designed to help educators, families, and communities in California to meet the challenge.

Achieving Balance Within Mathematics— Three Important Components

At the heart of mathematics is reasoning. One cannot do mathematics without reasoning. . . . Teachers need to provide their students with many opportunities to reason through their solutions, conjectures, and thinking processes. Opportunities in which very young students . . . make distinctions between irrelevant and relevant information or attributes, and justify relationships between sets can contribute to their ability to reason logically.

—S. Chapin, *The Partners in Change Handbook*

Mathematics education must provide students with a balanced instructional program. In such a program students become proficient in basic computational and procedural skills, develop conceptual understanding, and become adept at problem solving.

All three components are important; none is to be neglected or underemphasized. Balance, however, does not imply

One of the most important goals of mathematics is to teach students logical reasoning.

allocating set amounts of time for each of the three components. At some times students might be concentrating on lessons or tasks that focus on one component; at other times the focus may be on two or all three. As described in Chapter 4, "Instructional Strategies," different types of instruction seem to foster different components of mathematical competence. Nonetheless, recent studies suggest that all three components are interrelated (Geary 1994; Siegler and Stern 1998; Sophian 1997). For example, conceptual understanding provides important constraints on the types of procedures children use to solve mathematics problems; at the same time practicing procedures provides an opportunity to make inductions about the underlying concepts (Siegler and Stern 1998).

Balance Defined

Computational and procedural skills are those that all students should learn to use routinely and automatically. Students should practice basic computational and procedural skills sufficiently and use them frequently enough to commit them to memory. Frequent use is also required to ensure that these skills are retained and maintained over the years.

Mathematics makes sense to students who have a conceptual understanding of the domain. They know not only how to apply skills but also when to apply them and why they are being applied. They see the structure and logic of mathematics and use it flexibly, effectively, and appropriately. In seeing the larger picture and in understanding the underlying concepts, they are in a stronger position to apply their knowledge to new situations and problems and to recognize when they have made procedural errors.

Students who do not have a deep understanding of mathematics suspect that it is just a jumble of unrelated procedures and incomprehensible formulas. For example, children who do not understand the basic counting concepts view counting as a rote, a mechanical activity. They believe that the only correct way to count is by starting from left to right and by assigning each item a number (with a number name, such as "one") in succession (Briars and Siegler 1984). In contrast, children with a good conceptual understanding of counting understand that items can be counted in any order—starting from right to left, skipping around, and so forth—as long as each item is counted only once (Gelman and Meck 1983). A strong conceptual understanding of counting, in turn, provides the foundation for using increasingly sophisticated counting strategies to solve arithmetic problems (Geary, Bow-Thomas, and Yao 1992).

Problem solving in mathematics is a goal-related activity that involves applying skills, understandings, and experiences to resolve new, challenging, or perplexing mathematical situations. Problem solving involves a sequence of activities directed toward a specific mathematical goal, such as solving a word problem, a task that often involves the use of a series of mathematical procedures and a conceptual representation of the problem to be solved (Geary 1994; Siegler and Crowley 1994; Mayer 1985).

When students apply basic computational and procedural skills and understandings to solve new or perplexing problems, their basic skills are strengthened, the challenging problems they encounter can become routine, and their conceptual understanding deepens. They come to see mathematics as a way of finding solutions to problems that occur outside the classroom. Thus, students grow in their ability and persistence in problem solving through experience in solving problems at a variety of levels of difficulty and at every level in their mathematical development.

Basic Computational and Procedural Skills

For each level of mathematics, a specific set of basic computational and procedural skills must be learned. For example, students need to memorize the number facts of addition and multiplication of one-digit numbers and their corresponding subtraction and division facts. The ability to retrieve these facts automatically from long-term memory, in turn, makes the solving of more complex problems, such as multistep problems that involve basic arithmetic, quicker and less likely to result in errors (Geary and Widaman 1992). As students progress through elementary school, middle school, and high school, they should become proficient in the following skills:

- Finding correct answers to addition, subtraction, multiplication, and division problems
- Finding equivalencies for fractions, decimals, and percents
- Performing operations with fractions, decimals, and percents
- Measuring
- Finding perimeters and areas of simple figures
- Interpreting graphs encountered in daily life
- Finding the mean and median of a set of data from the real world
- Using scientific notation to represent very large or very small numbers
- Using basic geometry, including the Pythagorean theorem
- Finding the equation of a line, given two points through which it passes
- Solving linear equations and systems of linear equations

This list, which is by no means exhaustive, is provided for illustrative purposes only. Several factors should be considered in the development and maintenance of basic computational and procedural skills:

- Students must practice skills in order to become proficient. Practice should be varied and should be included both in homework assignments and in classroom activities. Teachers, students, and parents should realize that students must spend substantial time and exert significant effort to learn a skill and to maintain it for the long term (Ericsson, Krampe, and Tesch-Römer 1993).
- Basic computational and procedural skills develop over time, and they increase in depth and complexity through the years. For example, the ability to interpret information presented graphically begins at the primary level and extends to more sophisticated procedures as students progress through the grades.
- The development of basic computational and procedural skills requires that students be able to distinguish among different basic procedures by understanding what the procedures do. Only then will students have the basis for determining when to use the procedures they learn. For example, students must know the procedures involved in adding and multiplying fractions, and they must understand how and why these procedures produce different results.
- To maintain skills, students must use them frequently. Once students have learned to use the Pythagorean theorem, for example, they need to use it again and again in algebraic and geometric problems.
- Students may sometimes learn a skill more readily when they know how it

> **Chapter 1**
> Guiding Principles and Key Components of an Effective Mathematics Program
>
> The ability to recognize potential mathematical relationships is an important problem-solving technique, as is the identification of basic assumptions.

will be used or when they are intrigued by a problem that requires the skill.

Conceptual Understanding

Conceptual understanding is important at all levels of study. For example, during the elementary grades students should understand that:

- One way of thinking about multiplication is as repeated addition.
- One interpretation of fractions is as parts of a whole.
- Measurement of distances is fundamentally different from measurement of area.
- A larger sample generally provides more reliable information about the probability of an event than does a smaller sample.

As students progress through middle school and high school, they should, for example, understand that:

- The concepts of proportional relationships underlie similarity.
- The level sets of functions of two variables are curves in the coordinate plane.
- Factoring a polynomial function into irreducible factors helps locate the *x*-intercepts of its graph.
- Proofs are required to establish the truth of mathematical theorems.

Problem Solving

Problems occur in many forms. Some are simple and routine, providing practice for skill development. Others are more complex and take a longer time to complete. Whatever their nature, it is important that the kinds of problems students are asked to solve balance situations in the real world with more abstract situations. The process of solving problems generally has the following stages (Geary 1994; Mayer 1985):

- Formulation, analysis, and translation
- Integration and representation
- Solutions and justifications

Formulation, analysis, and translation. Problems may be stated in an imprecise form or in descriptions of puzzling or complex situations. The ability to recognize potential mathematical relationships is an important problem-solving technique, as is the identification of basic assumptions made directly or indirectly in the description of the situation, including extraneous or missing information. Important considerations in the formulation and analysis of any problem situation include determining mathematical hypotheses, making conjectures, recognizing existing patterns, searching for connections to known mathematical structures, and translating the gist of the problem into mathematical representations (e.g., equations).

Integration and representation. Important skills involved in the translation of a mathematical problem into a solvable equation are problem integration and representation. Integration involves putting together different pieces of information that are presented in complex problems, such as multistep problems. However such problems are represented, a wide variety of basic and technical skills are needed in solving problems; and, given this need, a mathematics program should include a substantial number of ready-to-solve exercises that are designed specifically to develop and reinforce such skills.

Solutions and justifications. Students should have a range of strategies to use in solving problems and should be encouraged to think about all possible procedures that might be used to aid in the solving of any particular problem, including but not limited to the following:

- Referring to and developing graphs, tables, diagrams, and sketches

- Computing
- Finding a simpler related problem
- Looking for patterns
- Estimating, conjecturing, and verifying
- Working backwards

Once the information in a complex problem has been integrated and translated into a mathematical representation, the student must be skilled at solving the associated equations and verifying the correctness of the solutions. Students might also identify relevant mathematical generalizations and seek connections to similar problems. From the earliest years students should be able to communicate and justify their solutions, starting with informal mathematical reasoning and advancing over the years to more formal mathematical proofs.

Connecting Skills, Conceptual Understanding, and Problem Solving

Basic computational and procedural skills, conceptual understanding, and problem solving form a web of mutually reinforcing elements in the curriculum. Computational and procedural skills are necessary for the actual solution of both simple and complex problems, and the practice of these skills provides a context for learning about the associated concepts and for discovering more sophisticated ways of solving problems (Siegler and Stern 1998). The development of conceptual understanding provides necessary constraints on the types of procedures students use to solve mathematics problems, enables students to detect when they have committed a procedural error, and facilitates the representation and translation phases of problem solving. Similarly, the process of applying skills in varying and increasingly complex problem-solving situations is one of the ways in which students not only refine their skills but also reinforce and strengthen their conceptual understanding and procedural competencies.

Key Components of an Effective Mathematics Program

Assumption: Proficiency is determined by student performance on valid and reliable measures aligned with the mathematics standards.

In an effective and well-designed mathematics program, students move steadily from what they already know to a mastery of skills, knowledge, and understanding. Their thinking progresses from an ability to explain what they are doing, to an ability to justify how and why they are doing it, to a stage at which they can derive formal proofs. The quality of instruction is a key factor in developing student proficiency in mathematics. In addition, several other factors or program components play an important role. They are discussed in the following section:

I. Assessment

Assessment should be the basis for instruction, and different types of assessment interact with the other components of an effective mathematics program.

In an effective mathematics program:

- Assessment is aligned with and guides instruction. Students are assessed frequently to determine whether they

are progressing steadily toward achieving the standards, and the results of this assessment are useful in determining instructional priorities and modifying curriculum and instruction. The assessment looks at the same balance (computational and procedural skills, conceptual understanding, and problem solving) emphasized in instruction.
- Assessments serve different purposes and are designed accordingly. Assessment for determining a student's placement in a mathematics program should cover a broad range of standards. These broad assessments measure whether or not students have prerequisite knowledge and allow them to demonstrate their full understanding of mathematics. Monitoring student progress daily or weekly requires a quick and focused measurement tool. Summative evaluation, which takes place at the end of a series of lessons or a course, provides specific and detailed information about which standards have or have not been achieved.
- Assessments are valid and reliable. A valid assessment measures the specific content it was designed to measure. An assessment instrument is reliable if it is relatively error-free and provides a stable result each time it is administered.
- Assessment can improve instruction when teachers use the results to analyze what students have learned and to reteach difficult concepts.

II. Instruction

The quality of instruction is the single most important component of an effective mathematics program. International comparisons show a high correlation between the quality of mathematics instruction and student achievement (Beaton et al. 1996).

In an effective mathematics program:

- Teachers possess an in-depth understanding of the content standards and the mathematics they are expected to teach and continually strive to increase their knowledge of content.
- Teachers are able to select research-based instructional strategies that are appropriate to the instructional goals and to students' needs.
- Teachers effectively organize instruction around goals that are tied to the standards and direct students' mathematical learning.
- Teachers use the results of assessment to guide instruction.

III. Instructional Time

Study after study has demonstrated the relationship between the time on task and student achievement (Stigler, Lee, and Stevenson 1987, 1283). Priority must be given to the teaching of mathematics, and instructional time must be protected from interruptions.

In an effective mathematics program:

- Adequate time is allocated to mathematics. Every day all students receive at least 50 to 60 minutes of mathematics instruction, not including homework. Additional instructional time is allocated for students who are, for whatever reason, performing substantially below grade level in mathematics. All students are encouraged to take mathematics courses throughout high school.
- Learning time is extended through homework that increases in complexity and duration as students mature. Homework should be valued and reviewed. The purpose of homework is to practice skills previously taught or to have students apply their previously learned knowledge and skills to new

problems. It should be assigned in amounts that are grade-level appropriate and, at least in the early grades, it should focus on independent practice and the application of skills already taught. For more advanced students, homework may be used as a means for exploring new concepts.
- During the great majority of allocated time, students are active participants in the instruction. *Active* can be described as the time during which students are engaged in thinking about mathematics or doing mathematics.
- Instructional time for mathematics is maximized and protected from such interruptions as calls to the office, public address announcements, and extracurricular activities.

IV. Instructional Resources

All teachers need high-quality instructional resources, but new teachers especially depend on well-designed resources and materials that are aligned with the standards.

In an effective mathematics program:
- Instructional resources focus on the grade-level standards. It may be necessary to go beyond the standards, however, both to provide meaningful enrichment and acceleration for fast learners and to introduce content needed for the mastery of standards in subsequent grades and courses. For example, the Algebra I standards do not mention complex numbers; yet quadratic equations, which often have complex roots, are fully developed in Algebra I. Therefore, an introduction to complex numbers may be included in Algebra I, both to avoid the artificial constraint of having all problems with real roots and to lay the foundation for the mastery of complex numbers in Algebra II.
- Instructional resources are factually and technically accurate and address the content outlined in the standards.
- Instructional resources emphasize depth of coverage. The most critical, highest-priority standards are addressed in the greatest depth. Ample practice is provided.
- Instructional resources are organized in a sequential, logical way. The resources are coordinated from level to level.
- Instructional options for teachers are included. For instance, a teacher's guide might explain the rationale and procedures for different ways of introducing a topic (e.g., through direct instruction or discovery-oriented instruction) and present various methods for assessing student progress. In addition to providing teachers with options, the resources should offer reliable guidelines for exercising those options.
- Resources balance basic computational and procedural skills, conceptual understanding, and problem solving and stress the interdependency of all three.
- Resources provide ample opportunities for students to explain their thinking, verbally and in writing, formally and informally.
- Resources supply ideas or tools for accommodating diverse student performance within any given classroom. They offer suggestions for reteaching a concept, providing additional practice for struggling students, or condensing instruction so that advanced students can concentrate on new material.

V. Instructional Grouping and Scheduling

Research shows that what students are taught has a greater effect on achievement

Chapter 1
Guiding Principles and Key Components of an Effective Mathematics Program

The primary management tool for teachers is the mathematics curriculum itself.

than does how they are grouped (Kulik 1992; Rogers 1991). The first focus of educators should always be on the quality of instruction. Grouping and scheduling are tools that educators can use to improve learning, not goals in and of themselves.

In an effective mathematics program:

- Grouping students according to their instructional needs improves student achievement (Benbow and Stanley 1996). An effective mathematics program (1) uses grouping options in accordance with variability within individual classrooms; and (2) maintains or changes grouping strategies in accordance with student performance on regular assessments.
- Cooperative group work is used judiciously, supplementing and expanding on initial instruction either delivered by teachers or facilitated through supervised exploration. Although students can often learn a great deal from one another and can benefit from the opportunity to discuss their thinking, the teacher is the primary leader in a class and maintains an active instructional role during cooperative learning. When cooperative group work is used, it should lead toward students' eventual independent demonstration of mastery of the standards and individual responsibility for learning.
- Cross-grade or cross-class grouping is an alternative to the more arbitrary practice of grouping according to chronological age or grade. Grouping by instructional needs across grade levels increases scheduling challenges for teachers and administrators near the beginning of a school year, but many teachers find the practice liberating later on because it reduces the number of levels for which a teacher must be prepared to teach in a single period.

VI. Classroom Management

Potentially, the primary management tool for teachers is the mathematics curriculum itself. When students are actively engaged in focused, rigorous mathematics, fewer opportunities for inappropriate behavior arise. When students are successful and their successes are made clear to them, they are more likely to become intrinsically motivated to work on mathematics.

In an effective mathematics program:

- Teachers are positive and optimistic about the prospect that all students can achieve. Research shows that teachers' self-esteem and enthusiasm for the subject matter have a greater effect on student achievement than does students' self-esteem (Clark 1997).
- Classrooms have a strong sense of purpose. Both academic and social expectations are clearly understood by teachers and students alike. Academic expectations relate directly to the standards.
- Intrinsic motivation is fostered by helping students to develop a deep understanding of mathematics, encouraging them to expend the effort needed to learn, and organizing instruction so that students experience satisfaction when they have mastered a difficult concept or skill. External reward systems are used sparingly; for example, as a temporary motivational device for older students who enter mathematics instruction without the intrinsic motivation to work hard.

VII. Professional Development

The preparation of teachers and support for their continuing professional development are critical to the quality of California schools. Research from other countries suggests that student achievement can improve when teachers are able to spend time together planning and evaluating instruction (Beaton et al. 1996).

In an effective mathematics program:

- Teachers have received excellent preservice training, are knowledgeable about mathematics content, and are able to use a wide variety of instructional strategies.
- Continuing teacher in-service training focuses on (1) enhancing teachers' proficiency in mathematics; and (2) providing pedagogical tools that help teachers to ensure that all students meet or exceed grade-level standards.
- Staff development is a long-term, planned investment strongly supported by the administration and designed to ensure that teachers continue to develop skills and knowledge in mathematics content and instructional options. "One-shot" staff development activities with no relationship to a long-term plan are recognized as having little lasting value.
- As with students, staff development actively engages teachers in mathematics and mathematics instruction. In addition to active involvement during classroom-style staff development, teachers have the opportunity to interact with students and staff developers during in-class coaching sessions.
- Individuals who have helped teachers bring their students to high achievement levels in mathematics are called on to demonstrate effective instructional practices with students.
- Teachers are given time and opportunities to work together to plan mathematics instruction. Districts and schools find creative ways to allow time for this planning.

VIII. Administrative Practices

Administrative support for mathematics instruction can help remind all those involved in education that reform efforts are not effective unless they contribute to increased achievement. Administrators can help teachers maintain a focus on high-quality instruction.

In an effective mathematics program:

- Mathematics achievement is among the highest priorities at the school.
- Long-term and short-term goals for the school, each grade level, and individuals are outlined clearly and reviewed frequently.
- Scheduling, grouping, and allocating personnel are shaped by a determination that all students will meet or exceed the mathematics standards.
- Principals demonstrate a strong sense of personal responsibility for achievement within their schools.
- Administrators consider using mathematics specialists to teach most or all of the mathematics classes or to coach other teachers.
- Administrators plan in advance for predictable contingencies, such as the need to realign instructional groups frequently, accommodate students transferring into the school, or redesign instruction for substantial numbers of students performing below grade level.
- Administrators and teachers collaborate on developing schoolwide management systems and schoolwide efforts to showcase mathematics for students, parents, and other members of the community.

IX. Community Involvement

Mathematics education is everybody's business. Parents, community members, and business and industry can all make significant contributions.

In an effective mathematics program:

- Parents are encouraged to be involved in education and are assisted in supporting their children's learning in mathematics. Parent comments are encouraged, valued, and used for program planning.
- Materials are organized so that parents, siblings, and community members can provide extended learning experiences.
- The community is used as a classroom that offers abundant examples of how and why mathematics is important in people's lives, work, and thinking.

The California Mathematics Content Standards

A high-quality mathematics program is essential for all students and provides every student with the opportunity to choose among the full range of future career paths. Mathematics, when taught well, is a subject of beauty and elegance, exciting in its logic and coherence. It trains the mind to be analytic—providing the foundation for intelligent and precise thinking.

To compete successfully in the worldwide economy, today's students must have a high degree of comprehension in mathematics. For too long schools have suffered from the notion that success in mathematics is the province of a talented few. Instead, a new expectation is needed: all students will attain California's mathematics academic content standards, and many will be inspired to achieve far beyond the minimum standards.

These content standards establish what every student in California can and needs to learn in mathematics. They are comparable to the standards of the most academically demanding nations, including Japan

Chapter 2
Mathematics Content Standards

The standards identify what all students in California public schools should know and be able to do at each grade level.

and Singapore—two high-performing countries in the Third International Mathematics and Science Study (TIMSS). Mathematics is critical for all students, not only those who will have careers that demand advanced mathematical preparation but all citizens who will be living in the twenty-first century. These standards are based on the premise that all students are capable of learning rigorous mathematics and learning it well, and all are capable of learning far more than is currently expected. Proficiency in most of mathematics is not an innate characteristic; it is achieved through persistence, effort, and practice on the part of students and rigorous and effective instruction on the part of teachers. Parents and teachers must provide support and encouragement.

The standards focus on essential content for all students and prepare students for the study of advanced mathematics, science and technical careers, and postsecondary study in all content areas. All students are required to grapple with solving problems; develop abstract, analytic thinking skills; learn to deal effectively and comfortably with variables and equations; and use mathematical notation effectively to model situations. The goal in mathematics education is for students to:

- Develop fluency in basic computational skills.
- Develop an understanding of mathematical concepts.
- Become mathematical problem solvers who can recognize and solve routine problems readily and can find ways to reach a solution or goal where no routine path is apparent.
- Communicate precisely about quantities, logical relationships, and unknown values through the use of signs, symbols, models, graphs, and mathematical terms.
- Reason mathematically by gathering data, analyzing evidence, and building arguments to support or refute hypotheses.
- Make connections among mathematical ideas and between mathematics and other disciplines.

The standards identify what all students in California public schools should know and be able to do at each grade level. Nevertheless, local flexibility is maintained with these standards. Topics may be introduced and taught at one or two grade levels before mastery is expected. Decisions about how best to teach the standards are left to teachers, schools, and school districts.

The standards emphasize computational and procedural skills, conceptual understanding, and problem solving. These three components of mathematics instruction and learning are not separate from each other; instead, they are intertwined and mutually reinforcing.

Basic, or computational and procedural, skills are those skills that all students should learn to use routinely and automatically. Students should practice basic skills sufficiently and frequently enough to commit them to memory.

Mathematics makes sense to students who have a conceptual understanding of the domain. They know not only *how* to apply skills but also *when* to apply them and *why* they should apply them. They understand the structure and logic of mathematics and use the concepts flexibly, effectively, and appropriately. In seeing the big picture and in understanding the concepts, they are in a stronger position to apply their knowledge to situations and problems they may not have encountered before and readily recognize when they have made procedural errors.

The mathematical reasoning standards are different from the other standards in that they do not represent a content domain. Mathematical reasoning is involved in all strands.

The standards do not specify how the curriculum should be delivered. Teachers may use direct instruction, explicit teaching, or knowledge-based discovery learning; investigatory, inquiry-based, problem-solving-based, guided discovery, set-theory-based, traditional, or progressive methods; or other ways in which to teach students the subject matter set forth in these standards. At the middle and high school levels, schools can use the standards with an integrated program or with the traditional course sequence of Algebra I, geometry, Algebra II, and so forth.

Schools that use these standards "enroll" students in a mathematical apprenticeship in which they practice skills, solve problems, apply mathematics to the real world, develop a capacity for abstract thinking, and ask and answer questions involving numbers or equations. Students need to know basic formulas, understand what they mean and why they work, and know when they should be applied. Students are also expected to struggle with thorny problems after learning to perform the simpler calculations on which they are based.

Teachers should guide students to think about why mathematics works in addition to how it works and should emphasize understanding of mathematical concepts as well as achievement of mathematical results. Students need to recognize that the solution to any given problem may be determined by employing more than one strategy and that the solution frequently raises new questions of its own: Does the answer make sense? Are there other, more efficient ways to arrive at the answer? Does the answer bring up more questions? Can I answer those? What other information do I need?

Problem solving involves applying skills, understanding, and experiences to resolve new or perplexing situations. It challenges students to apply their understanding of mathematical concepts in a new or complex situation, to exercise their computational and procedural skills, and to see mathematics as a way of finding answers to some of the problems that occur outside a classroom. Students grow in their ability and persistence in problem solving by extensive experience in solving problems at a variety of levels of difficulty and at every level in their mathematical development.

Problem solving, therefore, is an essential part of mathematics and is subsumed in every strand and in each of the disciplines in grades eight through twelve. Problem solving is not separate from content. Rather, students learn concepts and skills in order to apply them to solve problems in and outside school. Because problem solving is distinct from a content domain, its elements are consistent across grade levels.

The problems that students solve must address important mathematics. As students progress from grade to grade, they should deal with problems that (1) require increasingly more advanced knowledge and understanding of mathematics; (2) are increasingly complex

> Problem solving involves applying skills, understanding, and experiences to resolve new or perplexing situations.

(applications and purely mathematical investigations); and (3) require increased use of inductive and deductive reasoning and proof. In addition, problems should increasingly require students to make connections among mathematical ideas within a discipline and across domains. Each year students need to solve problems from all strands, although most of the problems should relate to the mathematics that students study that year. A good problem is one that is mathematically important; specifies the problem to be solved but not the solution path; and draws on grade-level appropriate skills and conceptual understanding.

Organization of the Standards

The mathematics content standards for kindergarten through grade seven are organized by grade level and are presented in five strands: Number Sense; Algebra and Functions; Measurement and Geometry; Statistics, Data Analysis, and Probability; and Mathematical Reasoning. Focus statements indicating the increasingly complex mathematical skills that will be required of students from kindergarten through grade seven are included at the beginning of each grade level; the statements indicate the ways in which the discrete skills and concepts form a cohesive whole. [The symbol ● indentifies the key standards to be covered in kindergarten through grade seven.]

The standards for grades eight through twelve are organized differently from those for kindergarten through grade seven. Strands are not used for organizational purposes because the mathematics studied in grades eight through twelve falls naturally under the discipline headings algebra, geometry, and so forth. Many schools teach this material in traditional courses; others teach it in an integrated program. To allow local educational agencies and teachers flexibility, the standards for grades eight through twelve do not mandate that a particular discipline be initiated and completed in a single grade. The content of these disciplines must be covered, and students enrolled in these disciplines are expected to achieve the standards regardless of the sequence of the disciplines.

Mathematics Standards and Technology

As rigorous mathematics standards are implemented for all students, the appropriate role of technology in the standards must be clearly understood. The following considerations may be used by schools and teachers to guide their decisions regarding mathematics and technology:

Students require a strong foundation in basic skills. Technology does not replace the need for all students to learn and master basic mathematics skills. All students must be able to add, subtract, multiply, and divide easily without the use of calculators or other electronic tools. In addition, all students need direct work and practice with the concepts and skills underlying the rigorous content described in the *Mathematics Content Standards for California Public Schools* so that they develop an understanding of quantitative concepts and relationships. The students' use of technology must build on these skills and understandings; it is not a substitute for them.

Technology should be used to promote mathematics learning. Technology can help promote students' understanding of mathematical concepts, quantitative reasoning, and achievement when used as a tool for solving problems, testing conjectures, accessing data, and verifying solutions. When students use electronic tools, databases, programming language, and simulations, they have opportunities to extend their comprehension, reasoning, and problem-solving skills beyond what is possible with traditional print resources. For example, graphing calculators allow students to see instantly the graphs of complex functions and to explore the impact of changes. Computer-based geometry construction tools allow students to see figures in three-dimensional space and experiment with the effects of transformations. Spreadsheet programs and databases allow students to key in data and produce various graphs as well as compile statistics. Students can determine the most appropriate ways to display data and quickly and easily make and test conjectures about the impact of change on the data set. In addition, students can exchange ideas and test hypotheses with a far wider audience through the Internet. Technology may also be used to reinforce basic skills through computer-assisted instruction, tutoring systems, and drill-and-practice software.

The focus must be on mathematics content. The focus must be on learning mathematics, using technology as a tool rather than as an end in itself. Technology makes more mathematics accessible and allows one to solve mathematical problems with speed and efficiency. However, technological tools cannot be used effectively without an understanding of mathematical skills, concepts, and relationships. As students learn to use electronic tools, they must also develop the quantitative reasoning necessary to make full use of those tools. They must also have opportunities to reinforce their estimation and mental math skills and the concept of place value so that they can quickly check their calculations for reasonableness and accuracy.

Technology is a powerful tool in mathematics. When used appropriately, technology may help students develop the skills, knowledge, and insight necessary to meet rigorous content standards in mathematics and make a successful transition to the world beyond school. The challenge for educators, parents, and policymakers is to ensure that technology supports, but is not a substitute for, the development of quantitative reasoning and problem-solving skills.

[Complete citations for the sources following some of the mathematics problems in this chapter appear in "Works Cited" in the references section. Many of the problems come from or are adapted from materials that are a part of the Third International Study of Mathematics and Science (TIMSS). TIMSS offers both a resource kit, *Attaining Excellence: A TIMSS Resource Kit*, and a Web site <http://www.csteep.bc.edu/TIMSS1/pubs_main.html>.]

Kindergarten Mathematics Content Standards

By the end of kindergarten, students understand small numbers, quantities, and simple shapes in their everyday environment. They count, compare, describe and sort objects, and develop a sense of properties and patterns.

Number Sense

1.0 Students understand the relationship between numbers and quantities (i.e., that a set of objects has the same number of objects in different situations regardless of its position or arrangement):

1.1 Compare two or more sets of objects (up to ten objects in each group) and identify which set is equal to, more than, or less than the other.

Are there more circles or more triangles in the following collection?

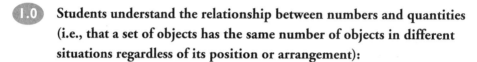

1.2 Count, recognize, represent, name, and order a number of objects (up to 30).

Which numbers are missing?

11, 12, 13, __, __, 16, 17, __, __, __, 21, 22, 23, 24.

1.3 Know that the larger numbers describe sets with more objects in them than the smaller numbers have.

2.0 Students understand and describe simple additions and subtractions:

2.1 Use concrete objects to determine the answers to addition and subtraction problems (for two numbers that are each less than 10).

Make as many pairs of numbers as possible by taking one number from each column so that each pair adds up to 10:

2	6
7	9
5	8
4	3
1	5
9	

Note: The sample problems illustrate the standards and are written to help clarify them. Some problems are written in a form that can be used directly with students; others will need to be modified, particularly in the primary grades, before they are used with students.

The symbol ● identifies the key standards for kindergarten.

Number Sense (Continued)

3.0 Students use estimation strategies in computation and problem solving that involve numbers that use the ones and tens places:

    3.1   Recognize when an estimate is reasonable.

Algebra and Functions

1.0 Students sort and classify objects:

    1.1   Identify, sort, and classify objects by attribute and identify objects that do not belong to a particular group (e.g., all these balls are green, those are red).

Students compare objects:

1. Which pencil is longer? Shorter?

2. Describe how the following 2 objects are the same or different.

3. Show students buttons sorted into 3 sets as shown and ask them to identify how buttons were sorted.

Measurement and Geometry

1.0 Students understand the concept of time and units to measure it; they understand that objects have properties, such as length, weight, and capacity, and that comparisons may be made by referring to those properties:

1.1 Compare the length, weight, and capacity of objects by making direct comparisons with reference objects (e.g., note which object is shorter, longer, taller, lighter, heavier, or holds more).

Who is the tallest girl in the class? The tallest boy?
Who is the oldest in the class?

1.2 Demonstrate an understanding of concepts of time (e.g., morning, afternoon, evening, today, yesterday, tomorrow, week, year) and tools that measure time (e.g., clock, calendar).

I left home at 9 o'clock in the morning and came back 2 hours later. What time did I come home?

1.3 Name the days of the week.

1.4 Identify the time (to the nearest hour) of everyday events (e.g., lunch time is 12 o'clock; bedtime is 8 o'clock at night).

2.0 Students identify common objects in their environment and describe the geometric features:

2.1 Identify and describe common geometric objects (e.g., circle, triangle, square, rectangle, cube, sphere, cone).

Which of these is a square?

Given 5 squares of the same size, can you make use of some or all of them to form a bigger square?

2.2 Compare familiar plane and solid objects by common attributes (e.g., position, shape, size, roundness, number of corners).

Statistics, Data Analysis, and Probability

1.0 **Students collect information about objects and events in their environment:**

    1.1   Pose information questions; collect data; and record the results using objects, pictures, and picture graphs.

    1.2   Identify, describe, and extend simple patterns (such as circles or triangles) by referring to their shapes, sizes, or colors.

Mathematical Reasoning

1.0 **Students make decisions about how to set up a problem:**

    1.1   Determine the approach, materials, and strategies to be used.

    1.2   Use tools and strategies, such as manipulatives or sketches, to model problems.

2.0 **Students solve problems in reasonable ways and justify their reasoning:**

    2.1   Explain the reasoning used with concrete objects and/or pictorial representations.

    2.2   Make precise calculations and check the validity of the results in the context of the problem.

          In a bag there are 4 apples, 3 oranges, 5 bananas, and 3 potatoes. How many pieces of fruit are in the bag altogether? How many different kinds of fruit are in the bag? How many objects altogether are in the bag?

Chapter 2 Mathematics Content Standards

Note: The sample problems illustrate the standards and are written to help clarify them. Some problems are written in a form that can be used directly with students; others will need to be modified, particularly in the primary grades, before they are used with students.

The symbol ● identifies the key standards for grade one.

Grade One Mathematics Content Standards

By the end of grade one, students understand and use the concept of ones and tens in the place value number system. Students add and subtract small numbers with ease. They measure with simple units and locate objects in space. They describe data and analyze and solve simple problems.

Number Sense

1.0 Students understand and use numbers up to 100:

1.1 Count, read, and write whole numbers to 100.

1.2 Compare and order whole numbers to 100 by using the symbols for less than, equal to, or greater than (<, =, >).

Which of the following are correct and which are incorrect?

(a) 75 > 76   (b) 48 < 42   (c) 89 > 91

(d) 59 < 67   (e) 34 = 33

1.3 Represent equivalent forms of the same number through the use of physical models, diagrams, and number expressions (to 20) (e.g., 8 may be represented as 4 + 4, 5 + 3, 2 + 2 + 2 + 2, 10 − 2, 11 − 3).

1.4 Count and group object in ones and tens (e.g., three groups of 10 and 4 equals 34, or 30 + 4).

A certain brand of chewing gum has 10 pieces in each pack. If there are 14 students, what is the smallest number of packs we must buy to make sure each student gets at least one piece of gum? If there are 19 students? What about 21 students?

There are 5 quarters, 9 dimes, 3 nickels, and 8 pennies. They are supposed to be put in piles of ten (coins). How many such piles can be formed by all these coins, and how many are left over?

1.5 Identify and know the value of coins and show different combinations of coins that equal the same value.

I have some pennies, nickels, and dimes in my pocket. I reach in and pull out three coins. How much money might I have? List all the possibilities.

Number Sense (Continued)

2.0 Students demonstrate the meaning of addition and subtraction and use these operations to solve problems:

- **2.1** Know the addition facts (sums to 20) and the corresponding subtraction facts and commit them to memory.

    I had 10 cupcakes, but I ate 3 of them. How many cupcakes do I have left? How many if I had 18 and ate 5?

- **2.2** Use the inverse relationship between addition and subtraction to solve problems.

- **2.3** Identify one more than, one less than, 10 more than, and 10 less than a given number.

- **2.4** Count by 2s, 5s, and 10s to 100.

    Which numbers are missing?

    24, 26, 28, 30, __, __, 36, __, 40, 42, 44, __, __, 50

    15, 20, 25, 30, __, __, 45, __, 55, 60, __, 70, __, 80

- **2.5** Show the meaning of addition (putting together, increasing) and subtraction (taking away, comparing, finding the difference).

- 2.6 Solve addition and subtraction problems with one- and two-digit numbers (e.g., 5 + 58 = __).

    Figure out how many pages I have read so far this week if I read 16 pages on Monday, 9 pages on Tuesday, none on Wednesday, and 14 pages on Thursday.

- 2.7 Find the sum of three one-digit numbers.

3.0 Students use estimation strategies in computation and problem solving that involve numbers that use the ones, tens, and hundreds places:

- 3.1 Make reasonable estimates when comparing larger or smaller numbers.

Algebra and Functions

1.0 Students use number sentences with operational symbols and expressions to solve problems:

   1.1 Write and solve number sentences from problem situations that express relationships involving addition and subtraction.

   Do the following problems in succession:
   1. Marie had some pencils in her desk. She put 5 more in her desk. Then she had 14. How many pencils did she have in her desk to start with?
   2. Eddie had 14 helium balloons. A number of them floated away. He had 5 left. How many did he lose?
   3. Nina had 14 seashells. That was 5 more than Pedro had. How many seashells did Pedro have?
   4. $5 + (\ ) = 6$? $(\ ) + 12 = 14$?

   1.2 Understand the meaning of the symbols $+$, $-$, $=$.

   1.3 Create problem situations that might lead to given number sentences involving addition and subtraction.

Measurement and Geometry

1.0 Students use direct comparison and nonstandard units to describe the measurements of objects:

   1.1 Compare the length, weight, and volume of two or more objects by using direct comparison or a nonstandard unit.

   Measure your desk by using the length of a ballpoint pen. How many ballpoint pens would be roughly equal to the length of your desk? The width of your desk? Which is longer?

   1.2 Tell time to the nearest half hour and relate time to events (e.g., before/after, shorter/longer).

2.0 Students identify common geometric figures, classify them by common attributes, and describe their relative position or their location in space:

   2.1 Identify, describe, and compare triangles, rectangles, squares, and circles, including the faces of three-dimensional objects.

   Make a picture of a house by using triangles, squares, and rectangles.

Measurement and Geometry (Continued)

2.2 Classify familiar plane and solid objects by common attributes, such as color, position, shape, size, roundness, or number of corners, and explain which attributes are being used for classification.

2.3 Give and follow directions about location.

> Here are pictures on a table of a ball, a girl, a horse, and a cat. Arrange them according to these directions:
> 1. Put the picture of the ball above the picture of the horse.
> 2. Put the picture of the girl on top of the picture of the horse.
> 3. Put the picture of the cat under the picture of the horse.

2.4 Arrange and describe objects in space by proximity, position, and direction (e.g., near, far, below, above, up, down, behind, in front of, next to, left or right of).

Statistics, Data Analysis, and Probability

1.0 **Students organize, represent, and compare data by category on simple graphs and charts:**

   1.1 Sort objects and data by common attributes and describe the categories.

   1.2 Represent and compare data (e.g., largest, smallest, most often, least often) by using pictures, bar graphs, tally charts, and picture graphs.

2.0 **Students sort objects and create and describe patterns by numbers, shapes, sizes, rhythms, or colors:**

   2.1 Describe, extend, and explain ways to get to a next element in simple repeating patterns (e.g., rhythmic, numeric, color, and shape).

Mathematical Reasoning

1.0 **Students make decisions about how to set up a problem:**

   1.1 Determine the approach, materials, and strategies to be used.

   1.2 Use tools, such as manipulatives or sketches, to model problems.

2.0 **Students solve problems and justify their reasoning:**

   2.1 Explain the reasoning used and justify the procedures selected.

   2.2 Make precise calculations and check the validity of the results from the context of the problem.

3.0 **Students note connections between one problem and another.**

Grade Two Mathematics Content Standards

By the end of grade two, students understand place value and number relationships in addition and subtraction, and they use simple concepts of multiplication. They measure quantities with appropriate units. They classify shapes and see relationships among them by paying attention to their geometric attributes. They collect and analyze data and verify the answers.

Number Sense

1.0 Students understand the relationship between numbers, quantities, and place value in whole numbers up to 1,000:

1.1 Count, read, and write whole numbers to 1,000 and identify the place value for each digit.

1.2 Use words, models, and expanded forms (e.g., 45 = 4 tens + 5) to represent numbers (to 1,000).

Kelly has 308 stickers. How many sets of hundreds, tens, and ones does she have?

1.3 Order and compare whole numbers to 1,000 by using the symbols <, =, >.

2.0 Students estimate, calculate, and solve problems involving addition and subtraction of two- and three-digit numbers:

2.1 Understand and use the inverse relationship between addition and subtraction (e.g., an opposite number sentence for 8 + 6 = 14 is 14 − 6 = 8) to solve problems and check solutions.

2.2 Find the sum or difference of two whole numbers up to three digits long.

Use drawings of tens and ones to help find the sum 37 + 17 and the difference 25 − 19.

2.3 Use mental arithmetic to find the sum or difference of two two-digit numbers.

In a game, Mysong and Naoki are making addition problems. They make two 2-digit numbers out of the four given numbers 1, 2, 3, and 4. Each number is used exactly once. The winner is the one who makes two numbers whose sum is the largest. Mysong had 43 and 21, while Naoki had 31 and 24. Who won the game? How do you know? Show how you can beat both Mysong and Naoki by making up two numbers with a larger sum than either (Adapted from TIMSS, gr. 4, V-4). (This problem also supports Mathematical Reasoning Standard 1.0.)

Note: The sample problems illustrate the standards and are written to help clarify them. Some problems are written in a form that can be used directly with students; others will need to be modified, particularly in the primary grades, before they are used with students.

The symbol ● identifies the key standards for grade two.

Number Sense (Continued)

3.0 **Students model and solve simple problems involving multiplication and division:**

- **3.1** Use repeated addition, arrays, and counting by multiples to do multiplication.

    Draw a simple picture of seating 30 people in rows of 10. Show and explain how this is related to multiplication. Do this also for rows of 3, and again for rows of 5.

- **3.2** Use repeated subtraction, equal sharing, and forming equal groups with remainders to do division.

    Kim decides to store away his marbles. He knows there are bags that hold up to 10 marbles in each. Kim has 38 marbles, and he tries to spend money on as few bags as he can. How many bags does he have to buy? How many if he has 51 marbles? (Keep in mind that there is no such thing as "half a bag" or "part of a bag.")

- **3.3** Know the multiplication tables of 2s, 5s, and 10s (to "times 10") and commit them to memory.

4.0 **Students understand that fractions and decimals may refer to parts of a set and parts of a whole:**

- **4.1** Recognize, name, and compare unit fractions from $\frac{1}{12}$ to $\frac{1}{2}$.

    True or false?

    1. One-fourth of a pie is larger than one-sixth of a pie.
    2. 1/4 > 1/3
    3. 1/7 < 1/9

- **4.2** Recognize fractions of a whole and parts of a group (e.g., one-fourth of a pie, two-thirds of 15 balls).

- **4.3** Know that when all fractional parts are included, such as four-fourths, the result is equal to the whole and to one.

5.0 **Students model and solve problems by representing, adding, and subtracting amounts of money:**

- **5.1** Solve problems using combinations of coins and bills.

    Lee has a wallet with 5 nickels, 9 dimes, and dollar bills. In how many ways can he pay with correct change for a pen worth $1.15? What about one worth 65 cents?

Number Sense (Continued)

5.2 Know and use the decimal notation and the dollar and cent symbols for money.

Which of the following show a correct use of symbols for money?

1. ¢32
2. 72¢
3. $1.25
4. 2.57$

6.0 Students use estimation strategies in computation and problem solving that involve numbers that use the ones, tens, hundreds, and thousands places:

6.1 Recognize when an estimate is reasonable in measurements (e.g., closest inch).

Algebra and Functions

1.0 Students model, represent, and interpret number relationships to create and solve problems involving addition and subtraction:

1.1 Use the commutative and associative rules to simplify mental calculations and to check results.

Draw pictures using dots to show:

1. Why $11 + 18 = 18 + 11$
2. Why $(11 + 5) + 17 = 11 + (5 + 17)$

If you know that $379 + 363 = 742$, what is the sum of $363 + 379$?

1.2 Relate problem situations to number sentences involving addition and subtraction.

Three classes at your school will see a play together in a large room. Room 1 has 18 students, Room 2 has 34 students, and Room 3 has 19 students. Figure out how many seats you will need. If Room 2 drops out but Room 4 with 29 students joins in, how many seats will you need then?

1.3 Solve addition and subtraction problems by using data from simple charts, picture graphs, and number sentences.

Measurement and Geometry

1.0 Students understand that measurement is accomplished by identifying a unit of measure, iterating (repeating) that unit, and comparing it to the item to be measured:

1.1 Measure the length of objects by iterating (repeating) a nonstandard or standard unit.

Four children measured the width of a room by counting how many paces it took them to cross it. It took Ana 9 paces, Erlane 8, Stephen 10, and Carlos 7. Who had the longest pace? (Adapted from TIMSS, gr. 4, L-8; gr. 8, L-12)

1.2 Use different units to measure the same object and predict whether the measure will be greater or smaller when a different unit is used.

Measure the length of your desk with a new crayon and with a new pencil. Which is greater, the number of crayon units or the number of pencil units?

1.3 Measure the length of an object to the nearest inch and/or centimeter.

1.4 Tell time to the nearest quarter hour and know relationships of time (e.g., minutes in an hour, days in a month, weeks in a year).

It took a bus 45 minutes to drive between the station and the bus barn. How long does it take to do four such trips nonstop?

Which is a longer period: 3 weeks or 19 days? 27 days or 4 weeks?

1.5 Determine the duration of intervals of time in hours (e.g., 11:00 a.m. to 4:00 p.m.).

2.0 Students identify and describe the attributes of common figures in the plane and of common objects in space:

2.1 Describe and classify plane and solid geometric shapes (e.g., circle, triangle, square, rectangle, sphere, pyramid, cube, rectangular prism) according to the number and shape of faces, edges, and vertices.

2.2 Put shapes together and take them apart to form other shapes (e.g., two congruent right triangles can be arranged to form a rectangle).

Statistics, Data Analysis, and Probability

1.0 **Students collect numerical data and record, organize, display, and interpret the data on bar graphs and other representations:**

1.1 Record numerical data in systematic ways, keeping track of what has been counted.

1.2 Represent the same data set in more than one way (e.g., bar graphs and charts with tallies).

1.3 Identify features of data sets (range and mode).

1.4 Ask and answer simple questions related to data representations.

2.0 **Students demonstrate an understanding of patterns and how patterns grow and describe them in general ways:**

2.1 Recognize, describe, and extend patterns and determine a next term in linear patterns (e.g., 4, 8, 12 . . . ; the number of ears on one horse, two horses, three horses, four horses).

> If there are two horses on a farm, how many horseshoes will we need to shoe all the horses? Show, in an organized way, how many horseshoes we will need for 3, 4, 5, 6, 7, 8, 9, and 10 horses.

2.2 Solve problems involving simple number patterns.

Mathematical Reasoning

1.0 Students make decisions about how to set up a problem:

1.1 Determine the approach, materials, and strategies to be used.

1.2 Use tools, such as manipulatives or sketches, to model problems.

2.0 Students solve problems and justify their reasoning:

2.1 Defend the reasoning used and justify the procedures selected.

2.2 Make precise calculations and check the validity of the results in the context of the problem.

3.0 Students note connections between one problem and another.

Grade Three   Mathematics Content Standards

By the end of grade three, students deepen their understanding of place value and their understanding of and skill with addition, subtraction, multiplication, and division of whole numbers. Students estimate, measure, and describe objects in space. They use patterns to help solve problems. They represent number relationships and conduct simple probability experiments.

Number Sense

1.0   Students understand the place value of whole numbers:

    1.1   Count, read, and write whole numbers to 10,000.

        What is the smallest whole number you can make using the digits 4, 3, 9, and 1? Use each digit exactly once (Adapted from TIMSS gr. 4, T-2).

    1.2   Compare and order whole numbers to 10,000.

    1.3   Identify the place value for each digit in numbers to 10,000.

    1.4   Round off numbers to 10,000 to the nearest ten, hundred, and thousand.

    1.5   Use expanded notation to represent numbers (e.g., $3{,}206 = 3{,}000 + 200 + 6$).

        True or false?

        $3{,}105 \times 3 = 9{,}000 + 300 + 10 + 5$

2.0   Students calculate and solve problems involving addition, subtraction, multiplication, and division:

    2.1   Find the sum or difference of two whole numbers between 0 and 10,000.

        1. $591 + 87 = ?$

        2. $1{,}283 + 6{,}074 = ?$

        3. $3{,}215 - 2{,}876 = ?$

        To prepare for recycling on Monday, Michael collected all the bottles in the house. He found 5 dark green ones, 8 clear ones with liquid still in them, 11 brown ones that used to hold root beer, 2 still with the cap on from his parents' cooking needs, and 4 more that were over-sized. How many bottles did Michael collect? (This problem also supports Mathematical Reasoning Standard 1.1.)

Note: The sample problems illustrate the standards and are written to help clarify them. Some problems are written in a form that can be used directly with students; others will need to be modified, particularly in the primary grades, before they are used with students.

The symbol ● identifies the key standards for grade three.

Number Sense (Continued)

2.2 Memorize to automaticity the multiplication table for numbers between 1 and 10.

2.3 Use the inverse relationship of multiplication and division to compute and check results.

2.4 Solve simple problems involving multiplication of multidigit numbers by one-digit numbers (3,671 × 3 = __).

A price list in a store states: pen sets, $3; magnets, $4; sticker sets, $6. How much would it cost to buy 5 pen sets, 7 magnets, and 8 sticker sets?

2.5 Solve division problems in which a multidigit number is evenly divided by a one-digit number (135 ÷ 5 = __).

2.6 Understand the special properties of 0 and 1 in multiplication and division.

True or false?

1. 24 × 0 = 24
2. 19 ÷ 1 = 19
3. 63 × 1 = 63
4. 0 ÷ 0 = 1

2.7 Determine the unit cost when given the total cost and number of units.

2.8 Solve problems that require two or more of the skills mentioned above.

A tree was planted 54 years before 1961. How old is the tree in 1998?

A class of 73 students go on a field trip. The school hires vans, each of which can seat a maximum of 10 students. The school policy is to seat as many students as possible in a van before using the next one. How many vans are needed?

3.0 Students understand the relationship between whole numbers, simple fractions, and decimals:

3.1 Compare fractions represented by drawings or concrete materials to show equivalency and to add and subtract simple fractions in context (e.g., ½ of a pizza is the same amount as 2/4 of another pizza that is the same size; show that 3/8 is larger than 1/4).

Which is longer, 1/3 of a foot or 5 inches? 2/3 of a foot or 9 inches?

Number Sense (Continued)

3.2 Add and subtract simple fractions (e.g., determine that ⅛ + ⅜ is the same as ½).

Find the values:

1. $\dfrac{1}{2} + \dfrac{3}{4} = ?$

2. $\dfrac{1}{2} - \dfrac{1}{8} = ?$

3.3 Solve problems involving addition, subtraction, multiplication, and division of money amounts in decimal notation and multiply and divide money amounts in decimal notation by using whole-number multipliers and divisors.

Pedro bought 5 pens, 2 erasers and 2 boxes of crayons. The pens cost 65 cents each, the erasers 25 cents each, and a box of crayons $1.10. The prices include tax, and Pedro paid with a ten-dollar bill. How much change did he get back?

3.4 Know and understand that fractions and decimals are two different representations of the same concept (e.g., 50 cents is ½ of a dollar, 75 cents is ¾ of a dollar).

Algebra and Functions

1.0 **Students select appropriate symbols, operations, and properties to represent, describe, simplify, and solve simple number relationships:**

1.1 Represent relationships of quantities in the form of mathematical expressions, equations, or inequalities.

1.2 Solve problems involving numeric equations or inequalities.

1.3 Select appropriate operational and relational symbols to make an expression true (e.g., if 4 __ 3 = 12, what operational symbol goes in the blank?).

1.4 Express simple unit conversions in symbolic form (e.g., __ inches = __ feet × 12).

If number of feet = number of yards × 3, and number of inches = number of feet × 12, how many inches are there in 4 yards?

Algebra and Functions (Continued)

1.5 Recognize and use the commutative and associative properties of multiplication (e.g., if $5 \times 7 = 35$, then what is 7×5? and if $5 \times 7 \times 3 = 105$, then what is $7 \times 3 \times 5$?).

When temperature is measured in both Celsius (C) and Fahrenheit (F), it is known that they are related by the following formula:
$9 \times C = (F - 32) \times 5$.

What is 50 degrees Fahrenheit in Celsius? (Note the explicit use of parentheses.)

2.0 **Students represent simple functional relationships:**

2.1 Solve simple problems involving a functional relationship between two quantities (e.g., find the total cost of multiple items given the cost per unit).

John wants to buy a dozen pencils. One store offers pencils at 6 for $1. Another offers them at 4 for 65 cents. Yet another sells pencils at 15 cents each. Where should John purchase his pencils in order to save the most money?

2.2 Extend and recognize a linear pattern by its rules (e.g., the number of legs on a given number of horses may be calculated by counting by 4s or by multiplying the number of horses by 4).

Here is the beginning of a pattern of tiles. Assuming that the pattern continues linearly, how many tiles will be in the sixth figure?
(Adapted from TIMSS gr. 4, K–6)

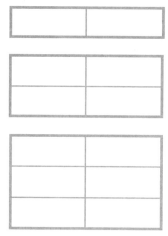

Measurement and Geometry

1.0 Students choose and use appropriate units and measurement tools to quantify the properties of objects:

    1.1   Choose the appropriate tools and units (metric and U.S.) and estimate and measure the length, liquid volume, and weight/mass of given objects.

    1.2   Estimate or determine the area and volume of solid figures by covering them with squares or by counting the number of cubes that would fill them.

            Which rectangle is NOT divided into four equal parts? (Adapted from TIMSS gr. 4, K-8)

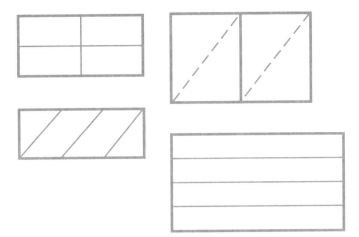

    1.3   Find the perimeter of a polygon with integer sides.

    1.4   Carry out simple unit conversions within a system of measurement (e.g., centimeters and meters, hours and minutes).

2.0 Students describe and compare the attributes of plane and solid geometric figures and use their understanding to show relationships and solve problems:

    2.1   Identify, describe, and classify polygons (including pentagons, hexagons, and octagons).

    2.2   Identify attributes of triangles (e.g., two equal sides for the isosceles triangle, three equal sides for the equilateral triangle, right angle for the right triangle).

    2.3   Identify attributes of quadrilaterals (e.g., parallel sides for the parallelogram, right angles for the rectangle, equal sides and right angles for the square).

Measurement and Geometry (Continued)

2.4 Identify right angles in geometric figures or in appropriate objects and determine whether other angles are greater or less than a right angle.

Which of the following triangles include an angle that is greater than a right angle?

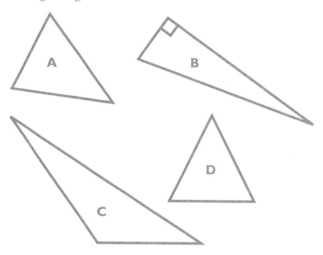

2.5 Identify, describe, and classify common three-dimensional geometric objects (e.g., cube, rectangular solid, sphere, prism, pyramid, cone, cylinder).

2.6 Identify common solid objects that are the components needed to make a more complex solid object.

Statistics, Data Analysis, and Probability

1.0 Students conduct simple probability experiments by determining the number of possible outcomes and make simple predictions:

1.1 Identify whether common events are certain, likely, unlikely, or improbable.

Are any of the following certain, likely, unlikely, or impossible?

1. Take two cubes each with the numbers 1, 2, 3, 4, 5, 6 written on its six faces. Throw them at random, and the sum of the numbers on the top faces is 12.

2. It snows on New Year's day.

3. A baseball game is played somewhere in this country on any Sunday in July.

4. It is sunny in June.

5. Pick any two one-digit numbers, and their sum is 17.

Statistics, Data Analysis, anad Probability (Continued)

- **1.2** Record the possible outcomes for a simple event (e.g., tossing a coin) and systematically keep track of the outcomes when the event is repeated many times.

- **1.3** Summarize and display the results of probability experiments in a clear and organized way (e.g., use a bar graph or a line plot).

- 1.4 Use the results of probability experiments to predict future events (e.g., use a line plot to predict the temperature forecast for the next day).

Mathematical Reasoning

1.0 **Students make decisions about how to approach problems:**

- 1.1 Analyze problems by identifying relationships, distinguishing relevant from irrelevant information, sequencing and prioritizing information, and observing patterns.

- 1.2 Determine when and how to break a problem into simpler parts.

2.0 **Students use strategies, skills, and concepts in finding solutions:**

- 2.1 Use estimation to verify the reasonableness of calculated results.

    Prove or disprove a classmate's claim that 49 is more than 21 because 9 is more than 1.

- 2.2 Apply strategies and results from simpler problems to more complex problems.

- 2.3 Use a variety of methods, such as words, numbers, symbols, charts, graphs, tables, diagrams, and models, to explain mathematical reasoning.

- 2.4 Express the solution clearly and logically by using the appropriate mathematical notation and terms and clear language; support solutions with evidence in both verbal and symbolic work.

- 2.5 Indicate the relative advantages of exact and approximate solutions to problems and give answers to a specified degree of accuracy.

- 2.6 Make precise calculations and check the validity of the results from the context of the problem.

Mathematical Reasoning (Continued)

3.0 Students move beyond a particular problem by generalizing to other situations:

3.1 Evaluate the reasonableness of the solution in the context of the original situation.

3.2 Note the method of deriving the solution and demonstrate a conceptual understanding of the derivation by solving similar problems.

3.3 Develop generalizations of the results obtained and apply them in other circumstances.

Grade Four | Mathematics Content Standards

By the end of grade four, students understand large numbers and addition, subtraction, multiplication, and division of whole numbers. They describe and compare simple fractions and decimals. They understand the properties of, and the relationships between, plane geometric figures. They collect, represent, and analyze data to answer questions.

Number Sense

1.0 Students understand the place value of whole numbers and decimals to two decimal places and how whole numbers and decimals relate to simple fractions. Students use the concepts of negative numbers:

1.1 Read and write whole numbers in the millions.

1.2 Order and compare whole numbers and decimals to two decimal places.

1.3 Round whole numbers through the millions to the nearest ten, hundred, thousand, ten thousand, or hundred thousand.

1.4 Decide when a rounded solution is called for and explain why such a solution may be appropriate.

Solve each of the following problems and observe the different roles played by the number 37 in each situation:

1. Four children shared 37 dollars equally. How much did each get?
2. Four children shared 37 pennies as equally as possible. How many pennies did each get?
3. Cars need to be rented for 37 children going on a field trip. Each car can take 12 children in addition to the driver. How many cars must be rented?

1.5 Explain different interpretations of fractions, for example, parts of a whole, parts of a set, and division of whole numbers by whole numbers; explain equivalents of fractions (see Standard 4.0).

Which number represents the shaded part of the figure? (Adapted from TIMSS gr. 4, M-5)

1. 2.8
2. 0.5
3. 0.2
4. 0.02

Chapter 2
Mathematics
Content
Standards

Note: The sample problems illustrate the standards and are written to help clarify them. Some problems are written in a form that can be used directly with students; others will need to be modified, particularly in the primary grades, before they are used with students.

The symbol ● identifies the key standards for grade four.

Number Sense (Continued)

True or false?

1. $1/4 > 2.54$
2. $5/2 < 2.6$
3. $12/18 = 2/3$
4. $4/5 < 13/15$ (Note the equivalence of fractions.)

1.6 Write tenths and hundredths in decimal and fraction notations and know the fraction and decimal equivalents for halves and fourths (e.g., ½ = 0.5 or 0.50; 7/4 = 1 ¾ = 1.75).

1.7 Write the fraction represented by a drawing of parts of a figure; represent a given fraction by using drawings; and relate a fraction to a simple decimal on a number line.

1.8 Use concepts of negative numbers (e.g., on a number line, in counting, in temperature, in "owing").

1.9 Identify on a number line the relative position of positive fractions, positive mixed numbers, and positive decimals to two decimal places.

True or false?

1. $-9 > -10$
2. $-31 < -29$

2.0 Students extend their use and understanding of whole numbers to the addition and subtraction of simple decimals:

2.1 Estimate and compute the sum or difference of whole numbers and positive decimals to two places.

Solve $55.73 - 48.25 = ?$

2.2 Round two-place decimals to one decimal or the nearest whole number and judge the reasonableness of the rounded answer.

Solve $17.91 + 2.18 = ?$

3.0 Students solve problems involving addition, subtraction, multiplication, and division of whole numbers and understand the relationships among the operations:

3.1 Demonstrate an understanding of, and the ability to use, standard algorithms for the addition and subtraction of multidigit numbers.

Solve $619,581 - 23,183 = ?$

Solve $6,747 + 321,105 = ?$

Number Sense (Continued)

3.2 Demonstrate an understanding of, and the ability to use, standard algorithms for multiplying a multidigit number by a two-digit number and for dividing a multidigit number by a one-digit number; use relationships between them to simplify computations and to check results.

Solve:

1. $783 \times 23 = ?$
2. $8{,}988/6 = ?$
3. $11{,}115/9 = ?$

3.3 Solve problems involving multiplication of multidigit numbers by two-digit numbers.

3.4 Solve problems involving division of multidigit numbers by one-digit numbers.

4.0 Students know how to factor small whole numbers:

4.1 Understand that many whole numbers break down in different ways (e.g., $12 = 4 \times 3 = 2 \times 6 = 2 \times 2 \times 3$).

In how many distinct ways can you write 60 as a product of two numbers?

4.2 Know that numbers such as 2, 3, 5, 7, and 11 do not have any factors except 1 and themselves and that such numbers are called prime numbers.

List all the distinct prime factors of 264.

Algebra and Functions

1.0 Students use and interpret variables, mathematical symbols, and properties to write and simplify expressions and sentences:

1.1 Use letters, boxes, or other symbols to stand for any number in simple expressions or equations (e.g., demonstrate an understanding and the use of the concept of a variable).

Tanya has read the first 78 pages of a 130-page book. Give the number of the sentence that can be used to find the number of pages Tanya must read to finish the book. (Adapted from TIMSS gr. 4, I-7)

1. $130 + 78 = \underline{}$
2. $\underline{} - 78 = 130$
3. $130 - 78 = \underline{}$
4. $130 - \underline{} = 178$

Algebra and Functions (Continued)

1.2 Interpret and evaluate mathematical expressions that now use parentheses.

Evaluate the two expressions: $(28 - 10) - 8 =$ ___ and $28 - (10 - 8) =$ ___.

1.3 Use parentheses to indicate which operation to perform first when writing expressions containing more than two terms and different operations.

Solve $(3 \times 12) - [(24/6) + 8] = ?$

Solve $([(18 + 31)/7] + 5) \times 9 = ?$

1.4 Use and interpret formulas (e.g., area = length × width or $A = lw$) to answer questions about quantities and their relationships.

There are many rules to get from Column A to Column B in the following table. Can you state one rule? (Adapted from TIMSS, gr. 4, J-5)

Column A	Column B
10	2
15	3
45	9
50	10

1.5 Understand that an equation such as $y = 3x + 5$ is a prescription for determining a second number when a first number is given.

2.0 **Students know how to manipulate equations:**

2.1 Know and understand that equals added to equals are equal.

2.2 Know and understand that equals multiplied by equals are equal.

Measurement and Geometry

1.0 Students understand perimeter and area:

   1.1   Measure the area of rectangular shapes by using appropriate units, such as square centimeter (cm²), square meter (m²), square kilometer (km²), square inch (in²), square yard (yd²), or square mile (mi²).

   Is the area of a 45×55 rectangle (in cm²) smaller or bigger than that of a square with the same perimeter?

   1.2   Recognize that rectangles that have the same area can have different perimeters.

   Draw a rectangle whose area is 1 and whose perimeter exceeds 50. Draw another rectangle with the same area whose perimeter exceeds 250.

   1.3   Understand that rectangles that have the same perimeter can have different areas.

   Draw a rectangle whose perimeter is 4 and whose area is less than 1/20.

   1.4   Understand and use formulas to solve problems involving perimeters and areas of rectangles and squares. Use those formulas to find the areas of more complex figures by dividing the figures into basic shapes.

   The length of a rectangle is 6 cm, and its perimeter is 16 cm. What is the area of the rectangle in square centimeters? (Adapted from TIMSS gr. 8, K–5)

2.0 Students use two-dimensional coordinate grids to represent points and graph lines and simple figures:

   2.1   Draw the points corresponding to linear relationships on graph paper (e.g., draw 10 points on the graph of the equation $y = 3x$ and connect them by using a straight line).

   2.2   Understand that the length of a horizontal line segment equals the difference of the x-coordinates.

   What is the length of the line segment joining the points (6, –4) and (21, –4)?

   2.3   Understand that the length of a vertical line segment equals the difference of the y-coordinates.

   What is the length of the line segment joining the points (121, 3) to (121, 17)?

Measurement and Geometry (Continued)

3.0 Students demonstrate an understanding of plane and solid geometric objects and use this knowledge to show relationships and solve problems:

3.1 Identify lines that are parallel and perpendicular.

3.2 Identify the radius and diameter of a circle.

3.3 Identify congruent figures.

3.4 Identify figures that have bilateral and rotational symmetry.

Let *AB*, *CD* be perpendicular diameters of a circle, as shown. If we reflect across the line segment *CD*, what happens to *A* and what happens to *B* under this reflection?

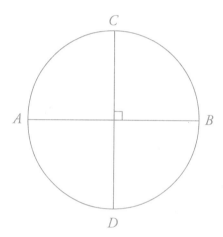

Craig folded a piece of paper in half and cut out a shape along the folded edge. Draw a picture to show what the cutout shape will look like when it is opened up and flattened out (Adapted from TIMSS gr. 4, T-5).

3.5 Know the definitions of a right angle, an acute angle, and an obtuse angle. Understand that 90°, 180°, 270°, and 360° are associated, respectively, with ¼, ½, ¾, and full turns.

3.6 Visualize, describe, and make models of geometric solids (e.g., prisms, pyramids) in terms of the number and shape of faces, edges, and vertices; interpret two-dimensional representations of three-dimensional objects; and draw patterns (of faces) for a solid that, when cut and folded, will make a model of the solid.

3.7 Know the definitions of different triangles (e.g., equilateral, isosceles, scalene) and identify their attributes.

Measurement and Geometry (Continued)

3.8 Know the definition of different quadrilaterals (e.g., rhombus, square, rectangle, parallelogram, trapezoid).

Explain which of the following statements are true and why.
1. All squares are rectangles.
2. All rectangles are squares.
3. All parallelograms are rectangles.
4. All rhombi are parallelograms.
5. Some parallelograms are squares.

Statistics, Data Analysis, and Probability

1.0 **Students organize, represent, and interpret numerical and categorical data and clearly communicate their findings:**

The following table shows the ages of the girls and boys in a club. Use the information in the table to complete the graph for ages 9 and 10. (Adapted from TIMSS gr. 4, S-1)

Ages	Number of Girls	Number of Boys
8	4	6
9	8	4
10	6	10

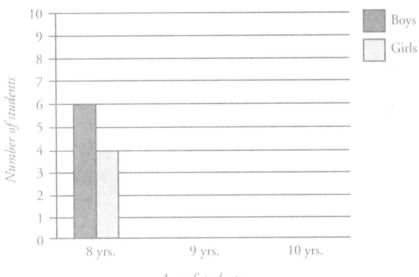

Ages of students

Statistics, Data Analysis, and Probability (Continued)

1.1 Formulate survey questions; systematically collect and represent data on a number line; and coordinate graphs, tables, and charts.

1.2 Identify the mode(s) for sets of categorical data and the mode(s), median, and any apparent outliers for numerical data sets.

1.3 Interpret one- and two-variable data graphs to answer questions about a situation.

2.0 **Students make predictions for simple probability situations:**

> Nine identical chips numbered 1 through 9 are put in a jar. When a chip is drawn from the jar, what is the probability that it has an even number? (Adapted from TIMSS gr. 8, N-18)

2.1 Represent all possible outcomes for a simple probability situation in an organized way (e.g., tables, grids, tree diagrams).

2.2 Express outcomes of experimental probability situations verbally and numerically (e.g., 3 out of 4; ¾).

Mathematical Reasoning

1.0 **Students make decisions about how to approach problems:**

1.1 Analyze problems by identifying relationships, distinguishing relevant from irrelevant information, sequencing and prioritizing information, and observing patterns.

1.2 Determine when and how to break a problem into simpler parts.

2.0 **Students use strategies, skills, and concepts in finding solutions:**

2.1 Use estimation to verify the reasonableness of calculated results.

2.2 Apply strategies and results from simpler problems to more complex problems.

2.3 Use a variety of methods, such as words, numbers, symbols, charts, graphs, tables, diagrams, and models, to explain mathematical reasoning.

2.4 Express the solution clearly and logically by using the appropriate mathematical notation and terms and clear language; support solutions with evidence in both verbal and symbolic work.

Mathematical Reasoning (Continued)

2.5 Indicate the relative advantages of exact and approximate solutions to problems and give answers to a specified degree of accuracy.

2.6 Make precise calculations and check the validity of the results from the context of the problem.

3.0 Students move beyond a particular problem by generalizing to other situations:

3.1 Evaluate the reasonableness of the solution in the context of the original situation.

3.2 Note the method of deriving the solution and demonstrate a conceptual understanding of the derivation by solving similar problems.

3.3 Develop generalizations of the results obtained and apply them in other circumstances.

Grade Five — Mathematics Content Standards

By the end of grade five, students increase their facility with the four basic arithmetic operations applied to fractions, decimals, and positive and negative numbers. They know and use common measuring units to determine length and area and know and use formulas to determine the volume of simple geometric figures. Students know the concept of angle measurement and use a protractor and compass to solve problems. They use grids, tables, graphs, and charts to record and analyze data.

Number Sense

1.0 Students compute with very large and very small numbers, positive integers, decimals, and fractions and understand the relationship between decimals, fractions, and percents. They understand the relative magnitudes of numbers:

1.1 Estimate, round, and manipulate very large (e.g., millions) and very small (e.g., thousandths) numbers.

1.2 Interpret percents as a part of a hundred; find decimal and percent equivalents for common fractions and explain why they represent the same value; compute a given percent of a whole number.

A test had 48 problems. Joe got 42 correct.

1. What percent were correct?
2. What percent were wrong?
3. If Moe got 93.75% correct, how many problems did he get correct?

1.3 Understand and compute positive integer powers of nonnegative integers; compute examples as repeated multiplication.

1. List all the factors of 48. List all the factors of 36. List the common factors.

2. Extend the tables shown below:

$2^4 = 16$ $10^4 = 10,000$

$2^3 = 8$ $10^3 = 1,000$

$2^2 = 4$ $10^2 = 100$

$2^1 = ?$ $10^1 = ?$

$2^0 = ?$ $10^0 = ?$

Number Sense (Continued)

1.4 Determine the prime factors of all numbers through 50 and write the numbers as the product of their prime factors by using exponents to show multiples of a factor (e.g., $24 = 2 \times 2 \times 2 \times 3 = 2^3 \times 3$).

1.5 Identify and represent on a number line decimals, fractions, mixed numbers, and positive and negative integers.

2.0 Students perform calculations and solve problems involving addition, subtraction, and simple multiplication and division of fractions and decimals:

2.1 Add, subtract, multiply, and divide with decimals; add with negative integers; subtract positive integers from negative integers; and verify the reasonableness of the results.

Determine the following numbers:

1. $-3(6 + (-7))$

2. $\dfrac{(-8 + (-12))}{(-2)}$

3. $\dfrac{(-5 + 3)(-9 + (-1))}{2}$

2.2 Demonstrate proficiency with division, including division with positive decimals and long division with multidigit divisors.

Find the quotient:

6 divided by .025

2.3 Solve simple problems, including ones arising in concrete situations, involving the addition and subtraction of fractions and mixed numbers (like and unlike denominators of 20 or less), and express answers in the simplest form.

2.4 Understand the concept of multiplication and division of fractions.

Given the following three pairs of fractions (⅜ and ⅙, 5¾ and 2⅓, 16 and 12⅞), find for each pair its:

1. Sum
2. Difference
3. Product
4. Quotient in simplest terms

2.5 Compute and perform simple multiplication and division of fractions and apply these procedures to solving problems.

Algebra and Functions

1.0 **Students use variables in simple expressions, compute the value of the expression for specific values of the variable, and plot and interpret the results:**

    1.1   Use information taken from a graph or equation to answer questions about a problem situation.

            Joe's sister Mary is twice as old as he is. Mary is 16. How old is Joe?

    1.2   Use a letter to represent an unknown number; write and evaluate simple algebraic expressions in one variable by substitution.

            $3x + 2 = 14$. What is x?

    1.3   Know and use the distributive property in equations and expressions with variables.

    1.4   Identify and graph ordered pairs in the four quadrants of the coordinate plane.

            Plot the points $(1, 2)$, $(-4, -3)$, $(12, -1)$, $(0, 4)$, $(-4, 0)$.

    1.5   Solve problems involving linear functions with integer values; write the equation; and graph the resulting ordered pairs of integers on a grid.

Measurement and Geometry

1.0 **Students understand and compute the volumes and areas of simple objects:**

    1.1   Derive and use the formula for the area of a triangle and of a parallelogram by comparing it with the formula for the area of a rectangle (i.e., two of the same triangles make a parallelogram with twice the area; a parallelogram is compared with a rectangle of the same area by pasting and cutting a right triangle on the parallelogram).

            Find the area and perimeter.

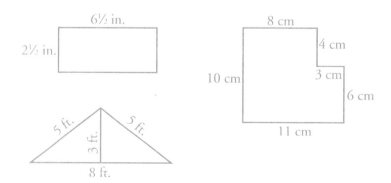

Measurement and Geometry (Continued)

- **1.2** Construct a cube and rectangular box from two-dimensional patterns and use these patterns to compute the surface area for these objects.

- **1.3** Understand the concept of volume and use the appropriate units in common measuring systems (i.e., cubic centimeter [cm^3], cubic meter [m^3], cubic inch [in.^{3}], cubic yard [yd.^{3}]) to compute the volume of rectangular solids.

- 1.4 Differentiate between, and use appropriate units of measures for, two- and three-dimensional objects (i.e., find the perimeter, area, volume).

2.0 **Students identify, describe, and classify the properties of, and the relationships between, plane and solid geometric figures:**

- **2.1** Measure, identify, and draw angles, perpendicular and parallel lines, rectangles, and triangles by using appropriate tools (e.g., straightedge, ruler, compass, protractor, drawing software).

- **2.2** Know that the sum of the angles of any triangle is 180° and the sum of the angles of any quadrilateral is 360° and use this information to solve problems.

- 2.3 Visualize and draw two-dimensional views of three-dimensional objects made from rectangular solids.

Statistics, Data Analysis, and Probability

1.0 **Students display, analyze, compare, and interpret different data sets, including data sets of different sizes:**

- 1.1 Know the concepts of mean, median, and mode; compute and compare simple examples to show that they may differ.

- 1.2 Organize and display single-variable data in appropriate graphs and representations (e.g., histogram, circle graphs) and explain which types of graphs are appropriate for various data sets.

- 1.3 Use fractions and percentages to compare data sets of different sizes.

- **1.4** Identify ordered pairs of data from a graph and interpret the meaning of the data in terms of the situation depicted by the graph.

- **1.5** Know how to write ordered pairs correctly; for example, (x, y).

Mathematical Reasoning

1.0 Students make decisions about how to approach problems:

1.1 Analyze problems by identifying relationships, distinguishing relevant from irrelevant information, sequencing and prioritizing information, and observing patterns.

1.2 Determine when and how to break a problem into simpler parts.

2.0 Students use strategies, skills, and concepts in finding solutions:

2.1 Use estimation to verify the reasonableness of calculated results.

2.2 Apply strategies and results from simpler problems to more complex problems.

2.3 Use a variety of methods, such as words, numbers, symbols, charts, graphs, tables, diagrams, and models, to explain mathematical reasoning.

2.4 Express the solution clearly and logically by using the appropriate mathematical notation and terms and clear language; support solutions with evidence in both verbal and symbolic work.

2.5 Indicate the relative advantages of exact and approximate solutions to problems and give answers to a specified degree of accuracy.

2.6 Make precise calculations and check the validity of the results from the context of the problem.

3.0 Students move beyond a particular problem by generalizing to other situations:

3.1 Evaluate the reasonableness of the solution in the context of the original situation.

3.2 Note the method of deriving the solution and demonstrate a conceptual understanding of the derivation by solving similar problems.

3.3 Develop generalizations of the results obtained and apply them in other circumstances.

Grade Six   Mathematics Content Standards

By the end of grade six, students have mastered the four arithmetic operations with whole numbers, positive fractions, positive decimals, and positive and negative integers; they accurately compute and solve problems. They apply their knowledge to statistics and probability. Students understand the concepts of mean, median, and mode of data sets and how to calculate the range. They analyze data and sampling processes for possible bias and misleading conclusions; they use addition and multiplication of fractions routinely to calculate the probabilities for compound events. Students conceptually understand and work with ratios and proportions; they compute percentages (e.g., tax, tips, interest). Students know about π and the formulas for the circumference and area of a circle. They use letters for numbers in formulas involving geometric shapes and in ratios to represent an unknown part of an expression. They solve one-step linear equations.

> **Note:** The sample problems illustrate the standards and are written to help clarify them. Some problems are written in a form that can be used directly with students; others will need to be modified, particularly in the primary grades, before they are used with students.
>
> The symbol ● identifies the key standards for grade six.

Number Sense

1.0 Students compare and order positive and negative fractions, decimals, a mixed numbers. Students solve problems involving fractions, ratios, pr tions, and percentages:

    1.1 Compare and order positive and negative fractions, decimals, and m numbers and place them on a number line.

    1.2 Interpret and use ratios in different contexts (e.g., batting averages, n hour) to show the relative sizes of two quantities, using appropriate r (a/b, a to b, $a:b$).

> Write the following as ratios:
> 1. The ratio of tricycles to tricycle wheels
> 2. The ratio of hands to fingers
> 3. If there are 6 tricycle wheels, how many tricycles are there?
> 4. If there are 45 fingers, how many hands are there?

    1.3 Use proportions to solve problems (e.g., determine the value of N if $4/7 = N/21$, find the length of a side of a polygon similar to a known polygon). Use cross-multiplication as a method for solving such problems, understanding it as the multiplication of both sides of an equation by a multiplicative inverse.

> Joe can type 9 words in 8 seconds. At this rate, how many words can he type in 2 minutes?

Number Sense (Continued)

1.4 Calculate given percentages of quantities and solve problems involving discounts at sales, interest earned, and tips.

2.0 **Students calculate and solve problems involving addition, subtraction, multiplication, and division:**

2.1 Solve problems involving addition, subtraction, multiplication, and division of positive fractions and explain why a particular operation was used for a given situation.

2.2 Explain the meaning of multiplication and division of positive fractions and perform the calculations (e.g., $5/8 \div 15/16 = 5/8 \times 16/15 = 2/3$).

Find n if:

1. $\dfrac{49}{21} = \dfrac{14}{n}$

2. $\dfrac{n}{3} = \dfrac{5}{7}$

(This problem also applies to Algebra and Functions Standard 1.1.)

2.3 Solve addition, subtraction, multiplication, and division problems, including those arising in concrete situations, that use positive and negative integers and combinations of these operations.

Simplify to make the calculation as simple as possible and identify the properties you used at each step:

1. $95 + 276 + 5$
2. $-19 + 37 + 19$
3. $-16(-8 + 9)$
4. $\left(-\dfrac{7}{8}\right)\left(\dfrac{17}{17}\right)$
5. $(-8)(-4)(19)(6 + (-6))$

2.4 Determine the least common multiple and the greatest common divisor of whole numbers; use them to solve problems with fractions (e.g., to find a common denominator to add two fractions or to find the reduced form for a fraction).

Algebra and Functions

1.0 Students write verbal expressions and sentences as algebraic expressions and equations; they evaluate algebraic expressions, solve simple linear equations, and graph and interpret their results:

    1.1 Write and solve one-step linear equations in one variable.

        $6y - 2 = 10$. What is y?

    1.2 Write and evaluate an algebraic expression for a given situation, using up to three variables.

    1.3 Apply algebraic order of operations and the commutative, associative, and distributive properties to evaluate expressions; and justify each step in the process.

    1.4 Solve problems manually by using the correct order of operations or by using a scientific calculator.

2.0 Students analyze and use tables, graphs, and rules to solve problems involving rates and proportions:

    2.1 Convert one unit of measurement to another (e.g., from feet to miles, from centimeters to inches).

        Suppose that one British pound is worth $1.50. In London a magazine costs 3 pounds. In San Francisco the same magazine costs $4.25. In which city is the magazine cheaper?

    2.2 Demonstrate an understanding that *rate* is a measure of one quantity per unit value of another quantity.

    2.3 Solve problems involving rates, average speed, distance, and time.

        Marcus took a train from San Francisco to San Jose, a distance of 54 miles. The train took 45 minutes for the trip. What was the average speed of the train?

3.0 Students investigate geometric patterns and describe them algebraically:

    3.1 Use variables in expressions describing geometric quantities (e.g., $P = 2w + 2l$, $A = \frac{1}{2}bh$, $C = \pi d$—the formulas for the perimeter of a rectangle, the area of a triangle, and the circumference of a circle, respectively).

        A rectangle has width w. Its length is one more than 3 times its width. Find the perimeter of the rectangle. (Your answer will be expressed in terms of w.)

    3.2 Express in symbolic form simple relationships arising from geometry.

Measurement and Geometry

1.0 Students deepen their understanding of the measurement of plane and solid shapes and use this understanding to solve problems:

 1.1 Understand the concept of a constant such as π; know the formulas for the circumference and area of a circle.

 1.2 Know common estimates of π (3.14; 22/7) and use these values to estimate and calculate the circumference and the area of circles; compare with actual measurements.

 1.3 Know and use the formulas for the volume of triangular prisms and cylinders (area of base × height); compare these formulas and explain the similarity between them and the formula for the volume of a rectangular solid.

 Find the volumes (dimensions are in cm).

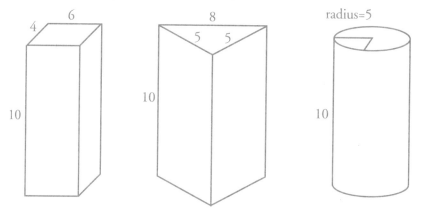

2.0 **Students identify and describe the properties of two-dimensional figures:**

 2.1 Identify angles as vertical, adjacent, complementary, or supplementary and provide descriptions of these terms.

 2.2 Use the properties of complementary and supplementary angles and the sum of the angles of a triangle to solve problems involving an unknown angle.

 Find the missing angles.

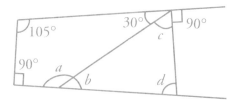

 2.3 Draw quadrilaterals and triangles from given information about them (e.g., a quadrilateral having equal sides but no right angles, a right isosceles triangle).

Statistics, Data Analysis, and Probability

1.0 **Students compute and analyze statistical measurements for data sets:**

    1.1   Compute the range, mean, median, and mode of data sets.

    1.2   Understand how additional data added to data sets may affect these computations of measures of central tendency.

    1.3   Understand how the inclusion or exclusion of outliers affects measures of central tendency.

    1.4   Know why a specific measure of central tendency (mean, median, mode) provides the most useful information in a given context.

2.0 **Students use data samples of a population and describe the characteristics and limitations of the samples:**

    2.1   Compare different samples of a population with the data from the entire population and identify a situation in which it makes sense to use a sample.

    2.2   Identify different ways of selecting a sample (e.g., convenience sampling, responses to a survey, random sampling) and which method makes a sample more representative for a population.

    2.3   Analyze data displays and explain why the way in which the question was asked might have influenced the results obtained and why the way in which the results were displayed might have influenced the conclusions reached.

    2.4   Identify data that represent sampling errors and explain why the sample (and the display) might be biased.

    2.5   Identify claims based on statistical data and, in simple cases, evaluate the validity of the claims.

3.0 **Students determine theoretical and experimental probabilities and use these to make predictions about events:**

    3.1   Represent all possible outcomes for compound events in an organized way (e.g., tables, grids, tree diagrams) and express the theoretical probability of each outcome.

    3.2   Use data to estimate the probability of future events (e.g., batting averages or number of accidents per mile driven).

    3.3   Represent probabilities as ratios, proportions, decimals between 0 and 1, and percentages between 0 and 100 and verify that the probabilities computed are reasonable; know that if P is the probability of an event, $1-P$ is the probability of an event not occurring.

Statistics, Data Analysis, and Probability (Continued)

3.4 Understand that the probability of either of two disjoint events occurring is the sum of the two individual probabilities and that the probability of one event following another, in independent trials, is the product of the two probabilities.

3.5 Understand the difference between independent and dependent events.

Mathematical Reasoning

1.0 **Students make decisions about how to approach problems:**

1.1 Analyze problems by identifying relationships, distinguishing relevant from irrelevant information, identifying missing information, sequencing and prioritizing information, and observing patterns.

1.2 Formulate and justify mathematical conjectures based on a general description of the mathematical question or problem posed.

1.3 Determine when and how to break a problem into simpler parts.

2.0 **Students use strategies, skills, and concepts in finding solutions:**

2.1 Use estimation to verify the reasonableness of calculated results.

2.2 Apply strategies and results from simpler problems to more complex problems.

2.3 Estimate unknown quantities graphically and solve for them by using logical reasoning and arithmetic and algebraic techniques.

2.4 Use a variety of methods, such as words, numbers, symbols, charts, graphs, tables, diagrams, and models, to explain mathematical reasoning.

2.5 Express the solution clearly and logically by using the appropriate mathematical notation and terms and clear language; support solutions with evidence in both verbal and symbolic work.

2.6 Indicate the relative advantages of exact and approximate solutions to problems and give answers to a specified degree of accuracy.

2.7 Make precise calculations and check the validity of the results from the context of the problem.

Mathematical Reasoning (Continued)

3.0 **Students move beyond a particular problem by generalizing to other situations:**

   3.1   Evaluate the reasonableness of the solution in the context of the original situation.

   3.2   Note the method of deriving the solution and demonstrate a conceptual understanding of the derivation by solving similar problems.

   3.3   Develop generalizations of the results obtained and the strategies used and apply them in new problem situations.

Grade Seven Mathematics Content Standards

By the end of grade seven, students are adept at manipulating numbers and equations and understand the general principles at work. Students understand and use factoring of numerators and denominators and properties of exponents. They know the Pythagorean theorem and solve problems in which they compute the length of an unknown side. Students know how to compute the surface area and volume of basic three-dimensional objects and understand how area and volume change with a change in scale. Students make conversions between different units of measurement. They know and use different representations of fractional numbers (fractions, decimals, and percents) and are proficient at changing from one to another. They increase their facility with ratio and proportion, compute percents of increase and decrease, and compute simple and compound interest. They graph linear functions and understand the idea of slope and its relation to ratio.

Number Sense

1.0 **Students know the properties of, and compute with, rational numbers expressed in a variety of forms:**

    1.1   Read, write, and compare rational numbers in scientific notation (positive and negative powers of 10) with approximate numbers using scientific notation.

           Write the following as a power of 10 or the product of a whole number and a power of 10:

           1. 10,000

           2. Ten billion

           3. 6,000,000

           4. 3 hundred thousand

    ● 1.2   Add, subtract, multiply, and divide rational numbers (integers, fractions, and terminating decimals) and take positive rational numbers to whole-number powers.

           Write the prime factorization of the following numbers:

           840   396   605   1,859

    1.3   Convert fractions to decimals and percents and use these representations in estimations, computations, and applications.

           Change to decimals:

           $\frac{7}{8}$ $\frac{7}{11}$

Number Sense (Continued)

1.4 Differentiate between rational and irrational numbers.

1.5 Know that every rational number is either a terminating or repeating decimal and be able to convert terminating decimals into reduced fractions.

Change to fractions:

0.27   0.272727

Find the period of the repeating part of $\frac{41}{13}$.

1.6 Calculate the percentage of increases and decreases of a quantity.

1.7 Solve problems that involve discounts, markups, commissions, and profit and compute simple and compound interest.

Joe borrows $800 at 10% interest compounded every six months. How much interest will there be in 4 years?

2.0 Students use exponents, powers, and roots and use exponents in working with fractions:

2.1 Understand negative whole-number exponents. Multiply and divide expressions involving exponents with a common base.

Continue the sequence:

$2^4 = 16$

$2^3 = 8$

$2^2 = 4$

$2^1 = 2$

$2^0 = 1$

$2^{-1} = ?$

$2^{-2} = ?$

2.2 Add and subtract fractions by using factoring to find common denominators.

Make use of prime factors to compute:

1. $\frac{2}{28} + \frac{1}{49}$

2. $\frac{-5}{63} + \left(\frac{-7}{99}\right)$

3. $\left(\frac{42}{22}\right)\left(\frac{75}{63}\right)$

Chapter 2
Mathematics
Content
Standards

Grade Seven

Number Sense (Continued)

2.3 Multiply, divide, and simplify rational numbers by using exponent rules.

2.4 Use the inverse relationship between raising to a power and extracting the root of a perfect square integer; for an integer that is not square, determine without a calculator the two integers between which its square root lies and explain why.

Find the edge of a square which has an area of 81.

2.5 Understand the meaning of the absolute value of a number; interpret the absolute value as the distance of the number from zero on a number line; and determine the absolute value of real numbers.

Algebra and Functions

1.0 Students express quantitative relationships by using algebraic terminology, expressions, equations, inequalities, and graphs:

1.1 Use variables and appropriate operations to write an expression, an equation, an inequality, or a system of equations or inequalities that represents a verbal description (e.g., three less than a number, half as large as area A).

Write the following verbal statements as algebraic expressions:

1. The square of a is increased by the sum of twice a and 3.
2. The product of ½ of a and 3 is decreased by the quotient of a divided by (-4).

1.2 Use the correct order of operations to evaluate algebraic expressions such as $3(2x + 5)^2$.

Given $x = (-2)$ and $y = 5$ evaluate:

1. $x^2 + 2x - 3$

2. $\dfrac{y(xy-7)}{10}$

Algebra and Functions (Continued)

1.3 Simplify numerical expressions by applying properties of rational numbers (e.g., identity, inverse, distributive, associative, commutative) and justify the process used.

Name the property illustrated by each of the following:

1. $x(y + -y) = x(0)$
2. $x(y + -y) = xy + x(-y)$
3. $x(y + -y) = (y + -y)(x)$
4. $x(y + -y) = x(-y + y)$
5. $x(y(1/y)) = x(1)$

1.4 Use algebraic terminology (e.g., variable, equation, term, coefficient, inequality, expression, constant) correctly.

1.5 Represent quantitative relationships graphically and interpret the meaning of a specific part of a graph in the situation represented by the graph.

2.0 Students interpret and evaluate expressions involving integer powers and simple roots:

2.1 Interpret positive whole-number powers as repeated multiplication and negative whole-number powers as repeated division or multiplication by the multiplicative inverse. Simplify and evaluate expressions that include exponents.

2.2 Multiply and divide monomials; extend the process of taking powers and extracting roots to monomials when the latter results in a monomial with an integer exponent.

3.0 Students graph and interpret linear and some nonlinear functions:

3.1 Graph functions of the form $y = nx^2$ and $y = nx^3$ and use in solving problems.

3.2 Plot the values from the volumes of three-dimensional shapes for various values of the edge lengths (e.g., cubes with varying edge lengths or a triangle prism with a fixed height and an equilateral triangle base of varying lengths).

3.3 Graph linear functions, noting that the vertical change (change in y-value) per unit of horizontal change (change in x-value) is always the same and know that the ratio ("rise over run") is called the slope of a graph.

3.4 Plot the values of quantities whose ratios are always the same (e.g., cost to the number of an item, feet to inches, circumference to diameter of a circle). Fit a line to the plot and understand that the slope of the line equals the quantities.

Algebra and Functions (Continued)

4.0 **Students solve simple linear equations and inequalities over the rational numbers:**

- 4.1 Solve two-step linear equations and inequalities in one variable over the rational numbers, interpret the solution or solutions in the context from which they arose, and verify the reasonableness of the results.

- 4.2 Solve multistep problems involving rate, average speed, distance, and time or a direct variation.

Measurement and Geometry

1.0 **Students choose appropriate units of measure and use ratios to convert within and between measurement systems to solve problems:**

- 1.1 Compare weights, capacities, geometric measures, times, and temperatures within and between measurement systems (e.g., miles per hour and feet per second, cubic inches to cubic centimeters).

    Convert the following:

    1. 80 miles/hr. = ? ft./sec.

    2. 20 oz./min. = ? qts./day

- 1.2 Construct and read drawings and models made to scale.

- 1.3 Use measures expressed as rates (e.g., speed, density) and measures expressed as products (e.g., person-days) to solve problems; check the units of the solutions; and use dimensional analysis to check the reasonableness of the answer.

2.0 **Students compute the perimeter, area, and volume of common geometric objects and use the results to find measures of less common objects. They know how perimeter, area, and volume are affected by changes of scale:**

- 2.1 Use formulas routinely for finding the perimeter and area of basic two-dimensional figures and the surface area and volume of basic three-dimensional figures, including rectangles, parallelograms, trapezoids, squares, triangles, circles, prisms, and cylinders.

- 2.2 Estimate and compute the area of more complex or irregular two- and three-dimensional figures by breaking the figures down into more basic geometric objects.

Measurement and Geometry (Continued)

2.3 Compute the length of the perimeter, the surface area of the faces, and the volume of a three-dimensional object built from rectangular solids. Understand that when the lengths of all dimensions are multiplied by a scale factor, the surface area is multiplied by the square of the scale factor and the volume is multiplied by the cube of the scale factor.

2.4 Relate the changes in measurement with a change of scale to the units used (e.g., square inches, cubic feet) and to conversions between units (1 square foot = 144 square inches or [1 ft²] = [144 in²], 1 cubic inch is approximately 16.38 cubic centimeters or [1 in³] = [16.38 cm³]).

3.0 **Students know the Pythagorean theorem and deepen their understanding of plane and solid geometric shapes by constructing figures that meet given conditions and by identifying attributes of figures:**

3.1 Identify and construct basic elements of geometric figures (e.g., altitudes, midpoints, diagonals, angle bisectors, and perpendicular bisectors; central angles, radii, diameters, and chords of circles) by using a compass and straightedge.

3.2 Understand and use coordinate graphs to plot simple figures, determine lengths and areas related to them, and determine their image under translations and reflections.

3.3 Know and understand the Pythagorean theorem and its converse and use it to find the length of the missing side of a right triangle and the lengths of other line segments and, in some situations, empirically verify the Pythagorean theorem by direct measurement.

> What is the side length of an isosceles right triangle with hypotenuse $\sqrt{72}$?

3.4 Demonstrate an understanding of conditions that indicate two geometrical figures are congruent and what congruence means about the relationships between the sides and angles of the two figures.

3.5 Construct two-dimensional patterns for three-dimensional models, such as cylinders, prisms, and cones.

Measurement and Geometry (Continued)

3.6 Identify elements of three-dimensional geometric objects (e.g., diagonals of rectangular solids) and describe how two or more objects are related in space (e.g., skew lines, the possible ways three planes might intersect).

True or false? If true, give an example. If false, explain why.

Two planes in three-dimensional space can:

1. Intersect in a line.

2. Intersect in a single point.

3. Have no intersection at all.

Statistics, Data Analysis, and Probability

1.0 **Students collect, organize, and represent data sets that have one or more variables and identify relationships among variables within a data set by hand and through the use of an electronic spreadsheet software program:**

1.1 Know various forms of display for data sets, including a stem-and-leaf plot or box-and-whisker plot; use the forms to display a single set of data or to compare two sets of data.

1.2 Represent two numerical variables on a scatterplot and informally describe how the data points are distributed and any apparent relationship that exists between the two variables (e.g., between time spent on homework and grade level).

1.3 Understand the meaning of, and be able to compute, the minimum, the lower quartile, the median, the upper quartile, and the maximum of a data set.

Mathematical Reasoning

1.0 **Students make decisions about how to approach problems:**

1.1 Analyze problems by identifying relationships, distinguishing relevant from irrelevant information, identifying missing information, sequencing and prioritizing information, and observing patterns.

1.2 Formulate and justify mathematical conjectures based on a general description of the mathematical question or problem posed.

1.3 Determine when and how to break a problem into simpler parts.

Mathematical Reasoning (Continued)

2.0 **Students use strategies, skills, and concepts in finding solutions:**

2.1 Use estimation to verify the reasonableness of calculated results.

2.2 Apply strategies and results from simpler problems to more complex problems.

2.3 Estimate unknown quantities graphically and solve for them by using logical reasoning and arithmetic and algebraic techniques.

2.4 Make and test conjectures by using both inductive and deductive reasoning.

2.5 Use a variety of methods, such as words, numbers, symbols, charts, graphs, tables, diagrams, and models, to explain mathematical reasoning.

2.6 Express the solution clearly and logically by using the appropriate mathematical notation and terms and clear language; support solutions with evidence in both verbal and symbolic work.

2.7 Indicate the relative advantages of exact and approximate solutions to problems and give answers to a specified degree of accuracy.

2.8 Make precise calculations and check the validity of the results from the context of the problem.

3.0 **Students determine a solution is complete and move beyond a particular problem by generalizing to other situations:**

3.1 Evaluate the reasonableness of the solution in the context of the original situation.

3.2 Note the method of deriving the solution and demonstrate a conceptual understanding of the derivation by solving similar problems.

3.3 Develop generalizations of the results obtained and the strategies used and apply them to new problem situations.

Introduction to Grades Eight Through Twelve

The standards for grades eight through twelve are organized differently from those for kindergarten through grade seven. In this section strands are not used for organizational purposes as they are in the elementary grades because the mathematics studied in grades eight through twelve falls naturally under discipline headings: algebra, geometry, and so forth. Many schools teach this material in traditional courses; others teach it in an integrated fashion. To allow local educational agencies and teachers flexibility in teaching the material, the standards for grades eight through twelve do not mandate that a particular discipline be initiated and completed in a single grade. The core content of these subjects must be covered; students are expected to achieve the standards however these subjects are sequenced.

Standards are provided for Algebra I, geometry, Algebra II, trigonometry, mathematical analysis, linear algebra, probability and statistics, advanced placement probability and statistics, and calculus. Many of the more advanced subjects are not taught in every middle school or high school. Moreover, schools and districts have different ways of combining the subject matter in these various disciplines. For example, many schools combine some trigonometry, mathematical analysis, and linear algebra to form a precalculus course. Some districts prefer offering trigonometry content with Algebra II.

Table 1, "Mathematics Disciplines, by Grade Level," reflects typical grade-level groupings of these disciplines in both integrated and traditional curricula. The lightly shaded region reflects the minimum requirement for mastery by all students. The dark shaded region depicts content that is typically considered elective but that should also be mastered by students who complete the other disciplines in the lower grade levels and continue the study of mathematics.

Many other combinations of these advanced subjects into courses are possible. What is described in this section are standards for the academic content by discipline; this document does not endorse a particular choice of structure for courses or a particular method of teaching the mathematical content.

When students delve deeply into mathematics, they gain not only conceptual understanding of mathematical principles but also knowledge of, and experience with, pure reasoning. One of the most important goals of mathematics is to teach students logical reasoning. The logical reasoning inherent in the study of mathematics allows for applications to a broad range of situations in which answers to practical problems can be found with accuracy.

By grade eight, students' mathematical sensitivity should be sharpened. Students need to start perceiving logical subtleties and appreciate the need for sound mathematical arguments before making conclusions. As students progress in the study of mathematics, they learn to distinguish between inductive and deductive reasoning; understand the meaning of logical implication; test general assertions;

Table 1. **Mathematics Disciplines, by Grade Level**

Disciplines	Grades				
	Eight	Nine	Ten	Eleven	Twelve
Algebra I					
Geometry					
Algebra II					
Probability and Statistics					
Trigonometry					
Linear Algebra					
Mathematical Analysis					
Advanced Placement Probability and Statistics					
Calculus					

realize that one counterexample is enough to show that a general assertion is false; understand conceptually that although a general assertion is true in a few cases, it is not true in all cases; distinguish between something being proven and a mere plausibility argument; and identify logical errors in chains of reasoning.

Mathematical reasoning and conceptual understanding are not separate from content; they are intrinsic to the mathematical discipline students master at more advanced levels.

Algebra I Mathematics Content Standards

Symbolic reasoning and calculations with symbols are central in algebra. Through the study of algebra, a student develops an understanding of the symbolic language of mathematics and the sciences. In addition, algebraic skills and concepts are developed and used in a wide variety of problem-solving situations.

1.0 Students identify and use the arithmetic properties of subsets of integers and rational, irrational, and real numbers, including closure properties for the four basic arithmetic operations where applicable:

    1.1 Students use properties of numbers to demonstrate whether assertions are true or false.

2.0 Students understand and use such operations as taking the opposite, finding the reciprocal, taking a root, and raising to a fractional power. They understand and use the rules of exponents.

Simplify $\left(x^3 y^{1/2}\right)^6 \sqrt{xy}$.

3.0 Students solve equations and inequalities involving absolute values.

4.0 Students simplify expressions before solving linear equations and inequalities in one variable, such as $3(2x-5) + 4(x-2) = 12$.

5.0 Students solve multistep problems, including word problems, involving linear equations and linear inequalities in one variable and provide justification for each step.

A-1 Pager Company charges a $25 set-up fee plus a $6.50 monthly charge. Cheaper Beeper charges $8 per month with no set-up fee. Set up an inequality to determine how long one would need to have the pager until the A-1 Pager plan would be the less expensive one.

6.0 Students graph a linear equation and compute the *x*- and *y*-intercepts (e.g., graph $2x + 6y = 4$). They are also able to sketch the region defined by linear inequality (e.g., they sketch the region defined by $2x + 6y < 4$).

Find inequalities whose simultaneous solution defines the region shown below:

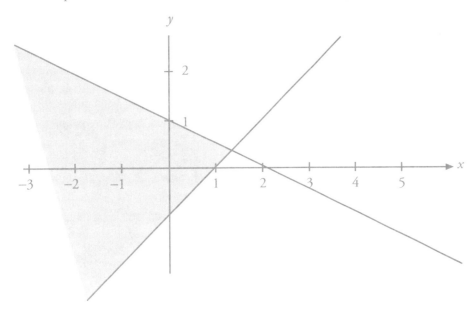

7.0 Students verify that a point lies on a line, given an equation of the line. Students are able to derive linear equations by using the point-slope formula.

Does the point (1, 2) lie on, above, or below the graph of the line $3x - 5y + 8 = 0$? Explain how you can be sure of your answer.

8.0 Students understand the concepts of parallel lines and perpendicular lines and how those slopes are related. Students are able to find the equation of a line perpendicular to a given line that passes through a given point.

9.0 Students solve a system of two linear equations in two variables algebraically and are able to interpret the answer graphically. Students are able to solve a system of two linear inequalities in two variables and to sketch the solution sets.

10.0 Students add, subtract, multiply, and divide monomials and polynomials. Students solve multistep problems, including word problems, by using these techniques.

Chapter 2
Mathematics Content Standards

Algebra I

11.0	Students apply basic factoring techniques to second- and simple third-degree polynomials. These techniques include finding a common factor for all terms in a polynomial, recognizing the difference of two squares, and recognizing perfect squares of binomials.
12.0	Students simplify fractions with polynomials in the numerator and denominator by factoring both and reducing them to the lowest terms.
13.0	Students add, subtract, multiply, and divide rational expressions and functions. Students solve both computationally and conceptually challenging problems by using these techniques.
	Solve for x and give a reason for each step:
	$\dfrac{2}{3x+1} + 2 = \dfrac{2}{3}$ (ICAS 1997)
14.0	Students solve a quadratic equation by factoring or completing the square.
15.0	Students apply algebraic techniques to solve rate problems, work problems, and percent mixture problems.
16.0	Students understand the concepts of a relation and a function, determine whether a given relation defines a function, and give pertinent information about given relations and functions.
17.0	Students determine the domain of independent variables and the range of dependent variables defined by a graph, a set of ordered pairs, or a symbolic expression.
18.0	Students determine whether a relation defined by a graph, a set of ordered pairs, or a symbolic expression is a function and justify the conclusion.
19.0	Students know the quadratic formula and are familiar with its proof by completing the square.

20.0	Students use the quadratic formula to find the roots of a second-degree polynomial and to solve quadratic equations.
	Suppose the graph of $y = px^2 + 5x + 2$ intersects the x-axis at two distinct points, where p is a constant. What are the possible values of p?

21.0	Students graph quadratic functions and know that their roots are the x-intercepts.

22.0	Students use the quadratic formula or factoring techniques or both to determine whether the graph of a quadratic function will intersect the x-axis in zero, one, or two points.

23.0	Students apply quadratic equations to physical problems, such as the motion of an object under the force of gravity.

24.0	Students use and know simple aspects of a logical argument:	
	24.1	Students explain the difference between inductive and deductive reasoning and identify and provide examples of each.
	24.2	Students identify the hypothesis and conclusion in logical deduction.
	24.3	Students use counterexamples to show that an assertion is false and recognize that a single counterexample is sufficient to refute an assertion.

25.0	Students use properties of the number system to judge the validity of results, to justify each step of a procedure, and to prove or disprove statements:	
	25.1	Students use properties of numbers to construct simple, valid arguments (direct and indirect) for, or formulate counterexamples to, claimed assertions.
	25.2	Students judge the validity of an argument according to whether the properties of the real number system and the order of operations have been applied correctly at each step.
	25.3	Given a specific algebraic statement involving linear, quadratic, or absolute value expressions or equations or inequalities, students determine whether the statement is true sometimes, always, or never.

Geometry | Mathematics Content Standards

The geometry skills and concepts developed in this discipline are useful to all students. Aside from learning these skills and concepts, students will develop their ability to construct formal, logical arguments and proofs in geometric settings and problems.

Note: The sample problems illustrate the standards and are written to help clarify them. Some problems are written in a form that can be used directly with students; others will need to be modified before they are used with students.

1.0 Students demonstrate understanding by identifying and giving examples of undefined terms, axioms, theorems, and inductive and deductive reasoning.

Using what you know about parallel lines cut by a transversal, show that the sum of the angles in a triangle is the same as the angle in a straight line, 180 degrees.

2.0 Students write geometric proofs, including proofs by contradiction.

If C is the center of the circle in the figure shown below, prove that angle b has twice the measure of angle a.

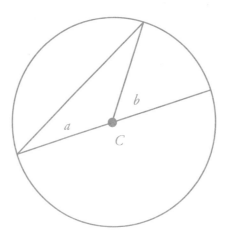

3.0 Students construct and judge the validity of a logical argument and give counterexamples to disprove a statement.

| 4.0 | Students prove basic theorems involving congruence and similarity. |

AB is a diameter of a circle centered at *O*. *CD* ⊥ *AB*. If the length of *AB* is 5, find the length of side *CD*. (CERT forthcoming)

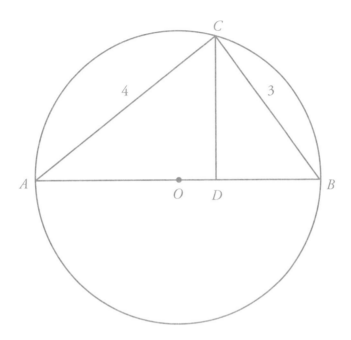

| 5.0 | Students prove that triangles are congruent or similar, and they are able to use the concept of corresponding parts of congruent triangles. |

| 6.0 | Students know and are able to use the triangle inequality theorem. |

| 7.0 | Students prove and use theorems involving the properties of parallel lines cut by a transversal, the properties of quadrilaterals, and the properties of circles. |

Prove that the figure formed by joining, in order, the midpoints of the sides of a quadrilateral is a parallelogram.

| 8.0 | Students know, derive, and solve problems involving the perimeter, circumference, area, volume, lateral area, and surface area of common geometric figures. |

| 9.0 | Students compute the volumes and surface areas of prisms, pyramids, cylinders, cones, and spheres; and students commit to memory the formulas for prisms, pyramids, and cylinders. |

Chapter 2
Mathematics
Content
Standards

Geometry

10.0 Students compute areas of polygons, including rectangles, scalene triangles, equilateral triangles, rhombi, parallelograms, and trapezoids.

The diagram below shows the overall floor plan for a house. It has right angles at three corners. What is the area of the house? What is the perimeter of the house? (CERT 1997)

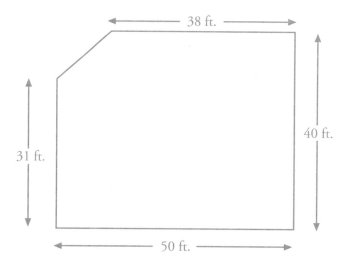

11.0 Students determine how changes in dimensions affect the perimeter, area, and volume of common geometric figures and solids.

A developer makes a scale model of the tract house that she is going to build. Each inch of distance on the scale model corresponds to 30 inches on the actual houses. If the scale model uses 2 square feet of roofing material, how much roofing material will be needed for one of the actual houses?

12.0 Students find and use measures of sides and of interior and exterior angles of triangles and polygons to classify figures and solve problems.

13.0 Students prove relationships between angles in polygons by using properties of complementary, supplementary, vertical, and exterior angles.

In the figure below, $\overline{AB} = \overline{BC} = \overline{CD}$. Find an expression for the measure of angle b in terms of the measure of angle a and prove that your expression is correct.

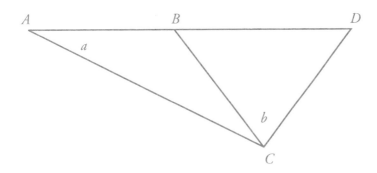

14.0 Students prove the Pythagorean theorem.

15.0 Students use the Pythagorean theorem to determine distance and find missing lengths of sides of right triangles.

16.0 Students perform basic constructions with a straightedge and compass, such as angle bisectors, perpendicular bisectors, and the line parallel to a given line through a point off the line.

17.0 Students prove theorems by using coordinate geometry, including the midpoint of a line segment, the distance formula, and various forms of equations of lines and circles.

18.0 Students know the definitions of the basic trigonometric functions defined by the angles of a right triangle. They also know and are able to use elementary relationships between them. For example, $\tan(x) = \sin(x)/\cos(x)$, $(\sin(x))^2 + (\cos(x))^2 = 1$.

Without using a calculator, determine which is larger, $\tan(60°)$ or $\tan(70°)$ and explain why.

19.0 Students use trigonometric functions to solve for an unknown length of a side of a right triangle, given an angle and a length of a side.

20.0 Students know and are able to use angle and side relationships in problems with special right triangles, such as 30°, 60°, and 90° triangles and 45°, 45°, and 90° triangles.

21.0 Students prove and solve problems regarding relationships among chords, secants, tangents, inscribed angles, and inscribed and circumscribed polygons of circles.

Use the perimeter of a regular hexagon inscribed in a circle to explain why $\pi > 3$. (ICAS 1997)

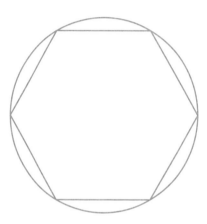

22.0 Students know the effect of rigid motions on figures in the coordinate plane and space, including rotations, translations, and reflections.

Algebra II   Mathematics Content Standards

This discipline complements and expands the mathematical content and concepts of Algebra I and geometry. Students who master Algebra II will gain experience with algebraic solutions of problems in various content areas, including the solution of systems of quadratic equations, logarithmic and exponential functions, the binomial theorem, and the complex number system.

Note: The sample problems illustrate the standards and are written to help clarify them. Some problems are written in a form that can be used directly with students; others will need to be modified before they are used with students.

1.0 Students solve equations and inequalities involving absolute value.

2.0 Students solve systems of linear equations and inequalities (in two or three variables) by substitution, with graphs, or with matrices.

Draw the region in the plane that is the solution set for the inequality $(x - 1)(x + 2y) > 0$.

3.0 Students are adept at operations on polynomials, including long division.

Divide $x^4 - 3x^2 + 3x$ by $x^2 + 2$.

Write the answer in the form: $\text{polynomial} + \dfrac{\text{linear polynomial}}{x^2 + 2}$.

4.0 Students factor polynomials representing the difference of squares, perfect square trinomials, and the sum and difference of two cubes.

5.0 Students demonstrate knowledge of how real and complex numbers are related both arithmetically and graphically. In particular, they can plot complex numbers as points in the plane.

6.0 Students add, subtract, multiply, and divide complex numbers.

7.0 Students add, subtract, multiply, divide, reduce, and evaluate rational expressions with monomial and polynomial denominators and simplify complicated rational expressions, including those with negative exponents in the denominator.

7.0 Simplify $\dfrac{\left(x^2 - x\right)^2}{x(x-1)^{-2}\left(x^2 + 3x - 4\right)}$.

8.0 Students solve and graph quadratic equations by factoring, completing the square, or using the quadratic formula. Students apply these techniques in solving word problems. They also solve quadratic equations in the complex number system.

In the figure shown below, the area between the two squares is 11 square inches. The sum of the perimeters of the two squares is 44 inches. Find the length of a side of the larger square. (ICAS 1997)

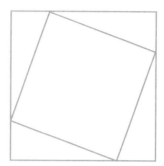

9.0 Students demonstrate and explain the effect that changing a coefficient has on the graph of quadratic functions; that is, students can determine how the graph of a parabola changes as a, b, and c vary in the equation $y = a(x-b)^2 + c$.

10.0 Students graph quadratic functions and determine the maxima, minima, and zeros of the function.

Find a quadratic function of x that has zeros at $x = -1$ and $x = 2$. Find a cubic equation of x that has zeros at $x = -1$ and $x = 2$ and nowhere else. (ICAS 1997)

11.0 Students prove simple laws of logarithms.

- **11.1** Students understand the inverse relationship between exponents and logarithms and use this relationship to solve problems involving logarithms and exponents.

- **11.2** Students judge the validity of an argument according to whether the properties of real numbers, exponents, and logarithms have been applied correctly at each step.

12.0 Students know the laws of fractional exponents, understand exponential functions, and use these functions in problems involving exponential growth and decay.

The number of bacteria in a colony was growing exponentially. At 1 p.m. yesterday the number of bacteria was 100, and at 3 p.m. yesterday it was 4,000. How many bacteria were there in the colony at 6 p.m. yesterday? (TIMSS)

13.0 Students use the definition of logarithms to translate between logarithms in any base.

14.0 Students understand and use the properties of logarithms to simplify logarithmic numeric expressions and to identify their approximate values.

Find the largest integer that is less than:

$\log_{10}(1,256)$

$\log_{10}(.029)$

15.0 Students determine whether a specific algebraic statement involving rational expressions, radical expressions, or logarithmic or exponential functions is sometimes true, always true, or never true.

16.0 Students demonstrate and explain how the geometry of the graph of a conic section (e.g., asymptotes, foci, eccentricity) depends on the coefficients of the quadratic equation representing it.

17.0 Given a quadratic equation of the form $ax^2 + by^2 + cx + dy + e = 0$, students can use the method for completing the square to put the equation into standard form and can recognize whether the graph of the equation is a circle, ellipse, parabola, or hyperbola. Students can then graph the equation.

Does the origin lie inside, outside, or on the geometric figure whose equation is $x^2 + y^2 - 10x + 10y - 1 = 0$? Explain your reasoning. (ICAS 1997)

18.0 Students use fundamental counting principles to compute combinations and permutations.

**Chapter 2
Mathematics Content Standards**

Algebra II

19.0	Students use combinations and permutations to compute probabilities.
20.0	Students know the binomial theorem and use it to expand binomial expressions that are raised to positive integer powers.
21.0	Students apply the method of mathematical induction to prove general statements about the positive integers.
22.0	Students find the general term and the sums of arithmetic series and of both finite and infinite geometric series.
23.0	Students derive the summation formulas for arithmetic series and for both finite and infinite geometric series.
24.0	Students solve problems involving functional concepts, such as composition, defining the inverse function and performing arithmetic operations on functions.

Which of the following functions are their own inverse functions? Use at least two different methods to answer this question and explain your methods:

$$f(x) = \frac{2}{x} \qquad g(x) = x^3 + 4 \qquad h(x) = \frac{2 + \ln x}{2 - \ln x} \qquad j(x) = \sqrt[3]{\frac{x^3 + 1}{x^3 - 1}}$$

(ICAS 1997)

25.0	Students use properties from number systems to justify steps in combining and simplifying functions.

Trigonometry — Mathematics Content Standards

Chapter 2
Mathematics Content Standards

Trigonometry uses the techniques that students have previously learned from the study of algebra and geometry. The trigonometric functions studied are defined geometrically rather than in terms of algebraic equations. Facility with these functions as well as the ability to prove basic identities regarding them is especially important for students intending to study calculus, more advanced mathematics, physics and other sciences, and engineering in college.

Note: The sample problems illustrate the standards and are written to help clarify them. Some problems are written in a form that can be used directly with students; others will need to be modified before they are used with students.

1.0 Students understand the notion of angle and how to measure it, in both degrees and radians. They can convert between degrees and radians.

2.0 Students know the definition of sine and cosine as y- and x-coordinates of points on the unit circle and are familiar with the graphs of the sine and cosine functions.

> Find an angle β between 0 and 2π such that $\cos(\beta) = \cos(6\pi/7)$ and $\sin(\beta) = -\sin(6\pi/7)$. Find an angle θ between 0 and 2π such that $\sin(\theta) = \cos(6\pi/7)$ and $\cos(\theta) = \sin(6\pi/7)$.

3.0 Students know the identity $\cos^2(x) + \sin^2(x) = 1$:

- **3.1** Students prove that this identity is equivalent to the Pythagorean theorem (i.e., students can prove this identity by using the Pythagorean theorem and, conversely, they can prove the Pythagorean theorem as a consequence of this identity).

- **3.2** Students prove other trigonometric identities and simplify others by using the identity $\cos^2(x) + \sin^2(x) = 1$. For example, students use this identity to prove that $\sec^2(x) = \tan^2(x) + 1$.

4.0 Students graph functions of the form $f(t) = A \sin(Bt + C)$ or $f(t) = A \cos(Bt + C)$ and interpret A, B, and C in terms of amplitude, frequency, period, and phase shift.

> On a graphing calculator, graph the function $f(x) = \sin(x)\cos(x)$. Select a window so that you can carefully examine the graph.
> 1. What is the apparent period of this function?
> 2. What is the apparent amplitude of this function?
> 3. Use this information to express f as a simpler trigonometric function.

Chapter 2
Mathematics
Content
Standards

Trigonometry

5.0	Students know the definitions of the tangent and cotangent functions and can graph them.
6.0	Students know the definitions of the secant and cosecant functions and can graph them.
7.0	Students know that the tangent of the angle that a line makes with the *x*-axis is equal to the slope of the line.
8.0	Students know the definitions of the inverse trigonometric functions and can graph the functions.
9.0	Students compute, by hand, the values of the trigonometric functions and the inverse trigonometric functions at various standard points.
10.0	Students demonstrate an understanding of the addition formulas for sines and cosines and their proofs and can use those formulas to prove and/or simplify other trigonometric identities. Use the addition formula for sine to find an expression for sin (75°).
11.0	Students demonstrate an understanding of half-angle and double-angle formulas for sines and cosines and can use those formulas to prove and/or simplify other trigonometric identities.
12.0	Students use trigonometry to determine unknown sides or angles in right triangles.
13.0	Students know the law of sines and the law of cosines and apply those laws to solve problems. A vertical pole sits between two points that are 60 feet apart. Guy wires to the top of that pole are staked at the two points. The guy wires are 40 feet and 35 feet long. How tall is the pole?
14.0	Students determine the area of a triangle, given one angle and the two adjacent sides.

15.0   Students are familiar with polar coordinates. In particular, they can determine polar coordinates of a point given in rectangular coordinates and vice versa.

16.0   Students represent equations given in rectangular coordinates in terms of polar coordinates.

         Express the circle of radius 2 centered at (2, 0) in polar coordinates.

17.0   Students are familiar with complex numbers. They can represent a complex number in polar form and know how to multiply complex numbers in their polar form.

18.0   Students know DeMoivre's theorem and can give nth roots of a complex number given in polar form.

19.0   Students are adept at using trigonometry in a variety of applications and word problems.

         A lighthouse stands 100 feet above the surface of the ocean. From what distance can it be seen? (Assume that the radius of the earth is 3,960 miles.)

Trigonometry

Chapter 2
Mathematics Content Standards

Mathematical Analysis — Mathematics Content Standards

Note: The sample problems illustrate the standards and are written to help clarify them. Some problems are written in a form that can be used directly with students; others will need to be modified before they are used with students.

This discipline combines many of the trigonometric, geometric, and algebraic techniques needed to prepare students for the study of calculus and strengthens their conceptual understanding of problems and mathematical reasoning in solving problems. These standards take a functional point of view toward those topics. The most significant new concept is that of limits. Mathematical analysis is often combined with a course in trigonometry or perhaps with one in linear algebra to make a yearlong precalculus course.

1.0 Students are familiar with, and can apply, polar coordinates and vectors in the plane. In particular, they can translate between polar and rectangular coordinates and can interpret polar coordinates and vectors graphically.

2.0 Students are adept at the arithmetic of complex numbers. They can use the trigonometric form of complex numbers and understand that a function of a complex variable can be viewed as a function of two real variables. They know the proof of DeMoivre's theorem.

3.0 Students can give proofs of various formulas by using the technique of mathematical induction.

Use mathematical induction to show that the sum of the interior angles in a convex polygon with n sides is $(n-2) \cdot 180°$.

4.0 Students know the statement of, and can apply, the fundamental theorem of algebra.

Find all cubic functions of x that have zeros at $x = -1$ and $x = 2$ and nowhere else. (ICAS 1997)

5.0 Students are familiar with conic sections, both analytically and geometrically:

  5.1 Students can take a quadratic equation in two variables; put it in standard form by completing the square and using rotations and translations, if necessary; determine what type of conic section the equation represents; and determine its geometric components (foci, asymptotes, and so forth).

	5.2	Students can take a geometric description of a conic section—for example, the locus of points whose sum of its distances from (1, 0) and (-1, 0) is 6—and derive a quadratic equation representing it.

6.0	Students find the roots and poles of a rational function and can graph the function and locate its asymptotes.

7.0	Students demonstrate an understanding of functions and equations defined parametrically and can graph them.

Sketch a graph of $f(x) = (x - 2)^2 - 1$. Sketch the graphs of $g(x) = f(|x|)$ and of $h(x) = |f(x)|$. Looking at your graph of $h(x)$, identify a value of x for which $h(x + 1) = h(x) - 3$.

8.0	Students are familiar with the notion of the limit of a sequence and the limit of a function as the independent variable approaches a number or infinity. They determine whether certain sequences converge or diverge.

Linear Algebra — Mathematics Content Standards

The general goal in this discipline is for students to learn the techniques of matrix manipulation so that they can solve systems of linear equations in any number of variables. Linear algebra is most often combined with another subject, such as trigonometry, mathematical analysis, or precalculus.

Note: The sample problems illustrate the standards and are written to help clarify them. Some problems are written in a form that can be used directly with students; others will need to be modified before they are used with students.

1.0 Students solve linear equations in any number of variables by using Gauss-Jordan elimination.

2.0 Students interpret linear systems as coefficient matrices and the Gauss-Jordan method as row operations on the coefficient matrix.

3.0 Students reduce rectangular matrices to row echelon form.

4.0 Students perform addition on matrices and vectors.

5.0 Students perform matrix multiplication and multiply vectors by matrices and by scalars.

6.0 Students demonstrate an understanding that linear systems are inconsistent (have no solutions), have exactly one solution, or have infinitely many solutions.

7.0 Students demonstrate an understanding of the geometric interpretation of vectors and vector addition (by means of parallelograms) in the plane and in three-dimensional space.

8.0 Students interpret geometrically the solution sets of systems of equations. For example, the solution set of a single linear equation in two variables is interpreted as a line in the plane, and the solution set of a two-by-two system is interpreted as the intersection of a pair of lines in the plane.

9.0 Students demonstrate an understanding of the notion of the inverse to a square matrix and apply that concept to solve systems of linear equations.

**Chapter 2
Mathematics
Content
Standards**

10.0 Students compute the determinants of 2 × 2 and 3 × 3 matrices and are familiar with their geometric interpretations as the area and volume of the parallelepipeds spanned by the images under the matrices of the standard basis vectors in two-dimensional and three-dimensional spaces.

11.0 Students know that a square matrix is invertible if, and only if, its determinant is nonzero. They can compute the inverse to 2 × 2 and 3 × 3 matrices using row reduction methods or Cramer's rule.

Linear Algebra

12.0 Students compute the scalar (dot) product of two vectors in n-dimensional space and know that perpendicular vectors have zero dot product.

Probability and Statistics | Mathematics Content Standards

This discipline is an introduction to the study of probability, interpretation of data, and fundamental statistical problem solving. Mastery of this academic content will provide students with a solid foundation in probability and facility in processing statistical information.

1.0 Students know the definition of the notion of *independent events* and can use the rules for addition, multiplication, and complementation to solve for probabilities of particular events in finite sample spaces.

2.0 Students know the definition of *conditional probability* and use it to solve for probabilities in finite sample spaces.

> A whole number between 1 and 30 is chosen at random. If the digits of the number that is chosen add up to 8, what is the probability that the number is greater than 12?

3.0 Students demonstrate an understanding of the notion of *discrete random variables* by using them to solve for the probabilities of outcomes, such as the probability of the occurrence of five heads in 14 coin tosses.

4.0 Students are familiar with the standard distributions (normal, binomial, and exponential) and can use them to solve for events in problems in which the distribution belongs to those families.

5.0 Students determine the mean and the standard deviation of a normally distributed random variable.

6.0 Students know the definitions of the *mean, median,* and *mode* of a distribution of data and can compute each in particular situations.

Chapter 2 Mathematics Content Standards

Note: The sample problems illustrate the standards and are written to help clarify them. Some problems are written in a form that can be used directly with students; others will need to be modified before they are used with students.

7.0 Students compute the variance and the standard deviation of a distribution of data.

Find the mean and standard deviation of the following seven numbers:

4   12   5   6   8   5   9

Make up another list of seven numbers with the same mean and a smaller standard deviation. Make up another list of seven numbers with the same mean and a larger standard deviation. (ICAS 1997)

8.0 Students organize and describe distributions of data by using a number of different methods, including frequency tables, histograms, standard line and bar graphs, stem-and-leaf displays, scatterplots, and box-and-whisker plots.

Advanced Placement Probability and Statistics

Mathematics Content Standards

This discipline is a technical and in-depth extension of probability and statistics. In particular, mastery of academic content for advanced placement gives students the background to succeed in the *Advanced Placement* examination in the subject.

1.0 Students solve probability problems with finite sample spaces by using the rules for addition, multiplication, and complementation for probability distributions and understand the simplifications that arise with independent events.

2.0 Students know the definition of *conditional probability* and use it to solve for probabilities in finite sample spaces.

> You have 5 coins in your pocket: 1 penny, 2 nickels, 1 dime, and 1 quarter. If you pull out 2 coins at random and they are collectively worth more than 10 cents, what is the probability that you pulled out a quarter?

3.0 Students demonstrate an understanding of the notion of *discrete random variables* by using this concept to solve for the probabilities of outcomes, such as the probability of the occurrence of five or fewer heads in 14 coin tosses.

4.0 Students understand the notion of a *continuous random variable* and can interpret the probability of an outcome as the area of a region under the graph of the probability density function associated with the random variable.

> Consider a continuous random variable X whose possible values are numbers between 0 and 2 and whose probability density function is given by $f(x) = 1 - \frac{1}{2}x$ for $0 \leq x \leq 2$. What is the probability that $X > 1$?

5.0 Students know the definition of the *mean of a discrete random variable* and can determine the mean for a particular discrete random variable.

Note: The sample problems illustrate the standards and are written to help clarify them. Some problems are written in a form that can be used directly with students; others will need to be modified before they are used with students.

6.0	Students know the definition of the *variance of a discrete random variable* and can determine the variance for a particular discrete random variable.
7.0	Students demonstrate an understanding of the standard distributions (normal, binomial, and exponential) and can use the distributions to solve for events in problems in which the distribution belongs to those families.
	Suppose that X is a normally distributed random variable with mean $\mu = 0$. If $P(X < c) = 2/3$, find $P(-c < X < c)$.
8.0	Students determine the mean and the standard deviation of a normally distributed random variable.
9.0	Students know the central limit theorem and can use it to obtain approximations for probabilities in problems of finite sample spaces in which the probabilities are distributed binomially.
10.0	Students know the definitions of the *mean, median,* and *mode of distribution* of data and can compute each of them in particular situations.
11.0	Students compute the variance and the standard deviation of a distribution of data.
12.0	Students find the line of best fit to a given distribution of data by using least squares regression.
13.0	Students know what the *correlation coefficient of two variables* means and are familiar with the coefficient's properties.
14.0	Students organize and describe distributions of data by using a number of different methods, including frequency tables, histograms, standard line graphs and bar graphs, stem-and-leaf displays, scatterplots, and box-and-whisker plots.
15.0	Students are familiar with the notions of a statistic of a distribution of values, of the sampling distribution of a statistic, and of the variability of a statistic.

Chapter 2
Mathematics Content Standards

Advanced Placement Probability and Statistics

16.0	Students know basic facts concerning the relation between the mean and the standard deviation of a sampling distribution and the mean and the standard deviation of the population distribution.
17.0	Students determine confidence intervals for a simple random sample from a normal distribution of data and determine the sample size required for a desired margin of error.
18.0	Students determine the *P*-value for a statistic for a simple random sample from a normal distribution.
19.0	Students are familiar with the *chi*-square distribution and *chi*-square test and understand their uses.

Calculus | Mathematics Content Standards

Chapter 2
Mathematics
Content
Standards

When taught in high school, calculus should be presented with the same level of depth and rigor as are entry-level college and university calculus courses. These standards outline a complete college curriculum in one variable calculus. Many high school programs may have insufficient time to cover all of the following content in a typical academic year. For example, some districts may treat differential equations lightly and spend substantial time on infinite sequences and series. Others may do the opposite. Consideration of the College Board syllabi for the Calculus AB and Calculus BC sections of the *Advanced Placement Examinations in Mathematics* may be helpful in making curricular decisions. Calculus is a widely applied area of mathematics and involves a beautiful intrinsic theory. Students mastering this content will be exposed to both aspects of the subject.

Note: The sample problems illustrate the standards and are written to help clarify them. Some problems are written in a form that can be used directly with students; others will need to be modified before they are used with students.

1.0 Students demonstrate knowledge of both the formal definition and the graphical interpretation of limit of values of functions. This knowledge includes one-sided limits, infinite limits, and limits at infinity. Students know the definition of convergence and divergence of a function as the domain variable approaches either a number or infinity:

  1.1 Students prove and use theorems evaluating the limits of sums, products, quotients, and composition of functions.

  1.2 Students use graphical calculators to verify and estimate limits.

  1.3 Students prove and use special limits, such as the limits of $(\sin(x))/x$ and $(1-\cos(x))/x$ as x tends to 0.

Evaluate the following limits, justifying each step:

$$\lim_{x \to 4} \frac{x-4}{\sqrt{x}-2}$$

$$\lim_{x \to 0} \frac{1-\cos(2x)}{\sin(3x)}$$

$$\lim_{x \to \infty} \left(x - \sqrt{x^2 - x} \right)$$

2.0 Students demonstrate knowledge of both the formal definition and the graphical interpretation of continuity of a function.

For what values of x is the function $f(x) = \dfrac{x^2 - 1}{x^2 - 4x + 3}$ continuous? Explain.

**Chapter 2
Mathematics
Content
Standards**

Calculus

3.0		Students demonstrate an understanding and the application of the intermediate value theorem and the extreme value theorem.
4.0		Students demonstrate an understanding of the formal definition of the derivative of a function at a point and the notion of differentiability:
	4.1	Students demonstrate an understanding of the derivative of a function as the slope of the tangent line to the graph of the function.
	4.2	Students demonstrate an understanding of the interpretation of the derivative as an instantaneous rate of change. Students can use derivatives to solve a variety of problems from physics, chemistry, economics, and so forth that involve the rate of change of a function.
	4.3	Students understand the relation between differentiability and continuity.
	4.4	Students derive derivative formulas and use them to find the derivatives of algebraic, trigonometric, inverse trigonometric, exponential, and logarithmic functions.

Find all points on the graph of $f(x) = \dfrac{x^2 - 2}{x + 1}$ where the tangent line is parallel to the tangent line at $x = 1$.

5.0	Students know the chain rule and its proof and applications to the calculation of the derivative of a variety of composite functions.
6.0	Students find the derivatives of parametrically defined functions and use implicit differentiation in a wide variety of problems in physics, chemistry, economics, and so forth.

For the curve given by the equation $\sqrt{x} + \sqrt{y} = 4$, use implicit differentiation to find $\dfrac{d^2 y}{dx^2}$.

7.0	Students compute derivatives of higher orders.
8.0	Students know and can apply Rolle's theorem, the mean value theorem, and L'Hôpital's rule.

9.0	Students use differentiation to sketch, by hand, graphs of functions. They can identify maxima, minima, inflection points, and intervals in which the function is increasing and decreasing.
10.0	Students know Newton's method for approximating the zeros of a function.
11.0	Students use differentiation to solve optimization (maximum-minimum problems) in a variety of pure and applied contexts.
	A man in a boat is 24 miles from a straight shore and wishes to reach a point 20 miles down shore. He can travel 5 miles per hour in the boat and 13 miles per hour on land. Find the minimal time for him to reach his destination and where along the shore he should land the boat to arrive as fast as possible.
12.0	Students use differentiation to solve related rate problems in a variety of pure and applied contexts.
13.0	Students know the definition of the definite integral by using Riemann sums. They use this definition to approximate integrals.
	The following is a Riemann sum that approximates the area under the graph of a function $f(x)$, between $x = a$ and $x = b$. Determine a possible formula for the function $f(x)$ and for the values of a and b:
	$$\sum_{i=1}^{n} \frac{2}{n} e^{1+\frac{2i}{n}}$$
14.0	Students apply the definition of the integral to model problems in physics, economics, and so forth, obtaining results in terms of integrals.
15.0	Students demonstrate knowledge and proof of the fundamental theorem of calculus and use it to interpret integrals as antiderivatives.
	If $f(x) = \int_{1}^{x} \sqrt{1+t^3}\, dt$, find $f''(2)$.
16.0	Students use definite integrals in problems involving area, velocity, acceleration, volume of a solid, area of a surface of revolution, length of a curve, and work.

Chapter 2
Mathematics Content Standards

Calculus

Chapter 2
Mathematics Content Standards

Calculus

17.0 Students compute, by hand, the integrals of a wide variety of functions by using techniques of integration, such as substitution, integration by parts, and trigonometric substitution. They can also combine these techniques when appropriate.

Evaluate the following:

$$\int \frac{\sin(1-\sqrt{x})}{\sqrt{x}} dx \qquad \int_1^e \frac{\ln x}{\sqrt{x}} dx \qquad \int_0^1 \sqrt{1+\sqrt{x}}\, dx$$

$$\int \arctan x\, dx \qquad \int \frac{\sqrt{x^2-1}}{x^3} dx \qquad \int \frac{dx}{e^x \sqrt{1-e^{2x}}}$$

18.0 Students know the definitions and properties of inverse trigonometric functions and the expression of these functions as indefinite integrals.

19.0 Students compute, by hand, the integrals of rational functions by combining the techniques in standard 17.0 with the algebraic techniques of partial fractions and completing the square.

20.0 Students compute the integrals of trigonometric functions by using the techniques noted above.

21.0 Students understand the algorithms involved in Simpson's rule and Newton's method. They use calculators or computers or both to approximate integrals numerically.

22.0 Students understand improper integrals as limits of definite integrals.

23.0 Students demonstrate an understanding of the definitions of convergence and divergence of sequences and series of real numbers. By using such tests as the comparison test, ratio test, and alternate series test, they can determine whether a series converges.

Determine whether the following alternating series converge absolutely, converge conditionally, or diverge:

$$\sum_{n=3}^{\infty} \frac{(-1)^n}{n \ln n} \qquad \sum_{n=3}^{\infty} (-1)^n \left(\frac{1+n}{n+\ln n} \right) \qquad \sum_{n=3}^{\infty} (-1)^n \left(\frac{2^n}{n!} \right)$$

24.0	Students understand and can compute the radius (interval) of the convergence of power series.
25.0	Students differentiate and integrate the terms of a power series in order to form new series from known ones.
26.0	Students calculate Taylor polynomials and Taylor series of basic functions, including the remainder term.
27.0	Students know the techniques of solution of selected elementary differential equations and their applications to a wide variety of situations, including growth-and-decay problems.

3

Grade-Level Considerations

Implementation of the standards will be challenging, especially during the early phases, when many students will not have the necessary foundational skills to master all of the expected grade-level mathematics content. This chapter provides a discussion of the mathematical considerations that went into the selection of the individual standards and describes the major roles some of them play in a standards-based curriculum. It also indicates areas where students may have difficulties, and, when possible, it provides techniques for easing them. Finally, it points out subtleties to which particular attention must be paid.

The chapter includes the following categories for each of the earlier grades:

- Areas of emphasis—Targets key areas of learning (These are taken directly from the *Mathematics Content Standards*.)
- Key standards—Identifies (●) some of the most important standards and tries to place them into context
- Elaboration—Provides added detail on these standards and on a number of related ones

- Grade-level readiness—Identifies areas of mathematics learning that are likely to present particular difficulties and concerns

In grades six and above, less attention is paid to the question of grade-level readiness, and proportionately more is paid to discussing the key standards and elaborations.

The five strands in the *Mathematics Content Standards* (Number Sense; Algebra and Functions; Measurement and Geometry; Statistics, Data Analysis, and Probability; and Mathematical Reasoning) organize information about the key standards for kindergarten through grade seven. **It should be noted that the strand of mathematical reasoning is different from the other four strands.** This strand, which is inherently embedded in each of the other strands, is fundamental in developing the basic skills and conceptual understanding for a solid mathematical foundation. It is important when looking at the standards to see the reasoning in all of them. Since this is the case, this chapter does not highlight key topics in the Mathematical Reasoning strand.

The section for grades eight through twelve in this chapter is organized by discipline, and only the basic ones—Algebra I; geometry; Algebra II; trigonometry; the precalculus course, mathematical analysis; and probability and statistics—are discussed in detail. The remaining courses are guided by other considerations, such as the *Advanced Placement (AP)* tests, and are outside the scope of this document.

The grade-level readiness information, which relates to difficult content areas in mathematics, is relevant to all teachers, students, and classrooms. This information will be particularly helpful during the transition to a standards-based mathematics curriculum.

The Strands

The content of the mathematics curriculum has frequently been divided into categories called strands. Like most systems of categories, the strands in mathematics were developed to break the content into a small set of manageable and understandable categories. Since there is no universal agreement on the selection of the parts, the use of strands is somewhat artificial; and many different systems have been suggested. In addition, it is often difficult to restrict a particular mathematical concept or skill to a single strand. Nonetheless, this framework continues the practice of presenting the content of mathematics in five strands for kindergarten through grade seven.

Because the content of mathematics builds and changes from grade to grade, the content in any one strand changes considerably over the course of mathematics programs for kindergarten through grade seven. Thus the strands serve only as an aid to organizing and thinking about the curriculum but no more than that. They describe the curriculum rather than define it. For the same reason the identification of strands does not mean that each is to be given equal weight in each year of mathematics education.

The general nature of each strand is described in the sections that follow.

Number Sense

Much of school mathematics depends on numbers, which are used to count, compute, measure, and estimate. The mathematics for this standard centers primarily on the development of number concepts; on computation with numbers (addition, subtraction, multiplication, division, finding powers and roots, and so forth); on numeration (systems for writing numbers, including base ten, fractions,

Mathematical reasoning is inherently embedded in each of the other strands.

negative numbers, rational numbers, percents, scientific notation, and so forth); and on estimation. At higher levels this strand includes the study of prime and composite numbers, of irrational numbers and their approximation by rationals, of real numbers, and of complex numbers.

Algebra and Functions

This strand involves two closely related subjects. Functions are rules that assign to each element in an initial set an element in a second set. For example, as early as kindergarten, children take collections of colored balls and sort them according to color, thereby assigning to each ball its color in the process. Later, students work with simple numeric functions, such as unit conversions that assign quantities of measurement; for example, 12 inches to each foot.

Functions are, therefore, one of the key areas of mathematical study. As indicated, they are encountered informally in the elementary grades and grow in prominence and importance with the student's increasing grasp of algebra in the higher grades. Beginning with the first year of algebra, functions are encountered at every turn.

Algebra proper again starts informally. It appears initially in its proper form in the third grade as "generalized arithmetic." In later grades algebra is the vital tool needed for solving equations and inequalities and using them as mathematical models of real situations. Students solve the problems that arise by translating from natural language—by which they communicate daily—to the abstract language of algebra and, conversely, from the formal language of algebra to natural language to demonstrate clear understanding of the concepts involved.

Functions are one of the key areas of mathematical study.

Measurement and Geometry

Geometry is the study of space and figures in space. In school any study of space, whether practical or theoretical, is put into the geometry strand. In the early grades this strand includes the use of measuring tools, such as rulers, and recognition of basic shapes, such as triangles, circles, squares, spheres, and cubes. In the later grades the content extends to the study of area and volume and the measurement of angles. In high school, plane geometry is studied both as an introduction to the concept of mathematical proof and as a fascinating structure that has profoundly influenced civilization for over 2,000 years.

Statistics, Data Analysis, and Probability

This strand includes the definitions and calculations of various averages and the analysis of data by classification and by graphical displays, taking into account randomness and bias in sampling. This strand has important connections with Strand 2, Algebra and Functions, and Strand 1, Number Sense, in the study of permutations and combinations and of Pascal's triangle. In the elementary grades effort is largely limited to collecting data and displaying it in graphs, in addition to calculating simple averages and performing probability experiments. This strand becomes more important in grade seven and above, when the students have gained the necessary skill with fractions and algebraic concepts in general so that statistics and their impact on daily life can be discussed with more sophistication than would have been possible earlier.

Mathematical Reasoning

Whenever a mathematical statement is justified, mathematical reasoning is involved. Mathematical reasoning in an inductive form appears in the early grades and is soon joined by deductive reasoning. Mathematical reasoning is involved in explaining arithmetic facts, in solving problems and puzzles at all levels, in understanding algorithms and formulas, and in justifying basic results in all areas of mathematics.

Mathematical reasoning, requiring careful, concise, and comprehensible proofs, is at the heart of mathematics and, indeed, is the essence of the discipline, differentiating it from others. Students must realize that assumptions are always involved in reaching conclusions, and they must recognize when assumptions are being made. Students must develop the habits of logical thinking and of recognizing and critically questioning all assumptions. In later life such reasoning skills will provide students with a foundation for making sound decisions and give them an invaluable defense against misleading claims.

Mathematical reasoning, requiring careful, concise, and comprehensible proofs, is at the heart of mathematics.

Key Standards

Number Sense	Algebra and Functions	Measurement and Geometry	Statistics, Data Analysis, and Probability	Mathematical Reasoning*
Kindergarten				
●1.0 1.1 1.2 1.3 2.0 ●2.1 3.0 3.1	1.0 ●1.1	●1.0 1.1 1.2 1.3 1.4 2.0 2.1 2.2	1.0 1.1 ●1.2	1.0 1.1 1.2 2.0 2.1 2.2
Grade One				
1.0 ●1.1 ●1.2 1.3 1.4 1.5 2.0 ●2.1 ●2.2 ●2.3 ●2.4 ●2.5 2.6 2.7 3.0 3.1	1.0 1.1 1.2 1.3	1.0 1.1 1.2 2.0 2.1 2.2 2.3 2.4	1.0 1.1 1.2 2.0 ●2.1	1.0 1.1 1.2 2.0 2.1 2.2 3.0
Grade Two				
1.0 ●1.1 1.2 ●1.3 2.0 ●2.1 ●2.2 2.3 ●3.0 ●3.1 ●3.2 ●3.3 4.0 ●4.1 ●4.2 ●4.3 5.0 ●5.1 ●5.2 6.0 6.1	1.0 ●1.1 1.2 1.3	1.0 1.1 1.2 ●1.3 1.4 1.5 ●2.0 ●2.1 ●2.2	●1.0 1.1 1.2 1.3 1.4 ●2.0 2.1 2.2	1.0 1.1 1.2 2.0 2.1 2.2 3.0

*It should be noted that the strand of mathematical reasoning is different from the other four strands. This strand, which is inherently embedded in each of the other strands, is fundamental in developing the basic skills and conceptual understanding for a solid mathematical foundation. It is important when looking at the standards to see the reasoning in all of them. Since this is the case, the key topics in the mathematical reasoning strand are not highlighted. Standards with the ● symbol are the most important ones to be covered within a grade level.

Key Standards (Continued)

Chapter 3
Grade-Level Considerations

Number Sense	Algebra and Functions	Measurement and Geometry	Statistics, Data Analysis, and Probability	Mathematical Reasoning
Grade Three				
1.0 1.1 1.2 **1.3** 1.4 **1.5** 2.0 **2.1 2.2 2.3 2.4** 2.5 2.6 2.7 2.8 3.0 3.1 **3.2 3.3** 3.4	1.0 **1.1** 1.2 1.3 1.4 1.5 2.0 **2.1** 2.2	1.0 1.1 **1.2 1.3** 1.4 2.0 **2.1 2.2 2.3** 2.4 2.5 2.6	1.0 1.1 **1.2 1.3** 1.4	1.0 1.1 1.2 2.0 2.1 2.2 2.3 2.4 2.5 2.6 3.0 3.1 3.2 3.3
Grade Four				
1.0 **1.1 1.2 1.3 1.4** 1.5 1.6 1.7 **1.8 1.9** 2.0 2.1 2.2 **3.0 3.1 3.2 3.3 3.4** 4.0 4.1 **4.2**	1.0 1.1 **1.2 1.3** 1.4 **1.5** **2.0 2.1 2.2**	1.0 1.1 1.2 1.3 1.4 **2.0 2.1 2.2 2.3** 3.0 3.1 3.2 3.3 3.4 3.5 3.6 3.7 3.8	1.0 1.1 1.2 1.3 2.0 2.1 2.2	1.0 1.1 1.2 2.0 2.1 2.2 2.3 2.4 2.5 2.6 3.0 3.1 3.2 3.3
Grade Five				
1.0 1.1 **1.2** 1.3 **1.4 1.5** 2.0 **2.1 2.2 2.3** 2.4 2.5	1.0 1.1 **1.2** 1.3 **1.4 1.5**	1.0 **1.1 1.2 1.3** 1.4 2.0 **2.1 2.2** 2.3	1.0 1.1 1.2 1.3 **1.4 1.5**	1.0 1.1 1.2 2.0 2.1 2.2 2.3 2.4 2.5 2.6 3.0 3.1 3.2 3.3
Grade Six				
1.0 1.1 1.2 1.3 1.4 **2.0** 2.1 2.2 **2.3 2.4**	1.0 **1.1** 1.2 1.3 1.4 2.0 2.1 **2.2** 2.3 3.0 3.1 3.2	1.0 **1.1** 1.2 1.3 2.0 2.1 **2.2** 2.3	1.0 1.1 1.2 1.3 1.4 2.0 2.1 **2.2 2.3 2.4 2.5** 3.0 **3.1** 3.2 **3.3** 3.4 **3.5**	1.0 1.1 1.2 1.3 2.0 2.1 2.2 2.3 2.4 2.5 2.6 2.7 3.0 3.1 3.2 3.3
Grade Seven				
1.0 1.1 **1.2** 1.3 **1.4 1.5** 1.6 **1.7** 2.0 2.1 **2.2 2.3** 2.4 **2.5**	1.0 1.1 1.2 **1.3** 1.4 1.5 2.0 2.1 2.2 3.0 3.1 3.2 **3.3** **3.4** **4.0 4.1 4.2**	1.0 1.1 1.2 **1.3** 2.0 2.1 2.2 2.3 2.4 3.0 3.1 3.2 **3.3** **3.4** 3.5 **3.6**	1.0 1.1 1.2 **1.3**	1.0 1.1 1.2 1.3 2.0 2.1 2.2 2.3 2.4 2.5 2.6 2.7 2.8 3.0 3.1 3.2 3.3

# Chapter 3: Grade-Level Considerations

Preface to Kindergarten Through Grade Seven

Mathematics, in the kindergarten through grade seven curriculum, starts with basic material and increases in scope and content as the years progress. It is like an inverted pyramid, with the entire weight of the developing subject resting on the core provided in kindergarten through grade two, when numbers, sets, and functions are introduced. If the introduction of the subject in the early grades is flawed, then later on, students can have extreme difficulty progressing; and their mathematical development can stop prematurely, leaving them, in one way or another, unable to fully realize their potential.

Because the teaching of mathematics in the early grades is largely synonymous with the problems given to the students, it is essential that students be presented with carefully constructed and mathematically accurate problems throughout their school careers. Problems which appear correct can actually be wrong, leading to serious misunderstandings on the part of the students. For example, the teacher might present the kindergarten standard for Algebra and Functions 1.1: "Identify, sort, and classify objects by attribute and identify objects that do not belong to a particular group." At first glance, the following exercise might seem appropriate for this standard:

A picture of three objects, a basketball, a bus, and a tennis ball, is shown to the students, and they are asked to tell which one does not belong.

This statement appears to present a perfectly reasonable problem. The difficulty is that, as stated, the question is not a problem in mathematics. From a mathematical point of view, the question is to determine which of these objects belongs to one set while the third belongs to a different one. It must be clear that unless the sets are specified in some way, the question cannot have a reasonable answer. In this case, the student must *guess* that the teacher is asking the student to sort objects by shape. The following might be asked instead: *We want to collect balls. Which of these objects should we select?* Or perhaps the contrapositive, *Which of these objects should not be included?* Another approach is to add colors; for example, coloring the bus and tennis ball blue and the basketball brown. Then a different question might be asked: *We want blue things. Which of these objects do we want?* or *We want round, blue objects. Which of these do we want?* But a question in the mathematics part of the curriculum should not be asked when the assumptions underlying what is wanted are not clearly stated.

In another example, the standard for Statistics, Data Analysis, and Probability 1.2 asks students to identify, describe, and

> It is essential that students be presented with carefully constructed and mathematically accurate problems throughout their school careers.

extend simple patterns involving shape, size, or color, such as a circle or triangle or red or blue. A possible problem illustrating the standard follows:

> The students are given a picture that shows in succession a rectangle, triangle, square, rectangle, triangle, square, blank, triangle, square. The students are asked to fill in the blank.

While this problem may seem to be a reasonable one (and an example of problems that all too commonly appear in the mathematics curricula of the lower grades), it cannot be considered a problem in mathematics. From a mathematical point of view, there is no correct answer to this problem unless more data are supplied to the students. Mathematics is about drawing logical conclusions from explicitly stated hypotheses. *Because there is no statement about the nature of the pattern in this case (e.g., does the pattern repeat itself every three terms? every seven terms? every nine terms?), students can only guess at what should be in the blank spot.*

The intent of the problem was probably to ask students to infer from the given data that the pattern, in all likelihood, repeats itself every three terms, leaving students to assume that a rectangle belongs in the blank spot. But if students were to start thinking that every mathematical situation always contains a hidden agenda for them to guess correctly before they can proceed, then both the teaching and learning of mathematics would be tremendously compromised. Observations from some university-level mathematicians suggest that this outcome may have already occurred with some students. Students' reluctance to take mathematical statements at face value has become a major stumbling block.

In an attempt to make mathematics "more relevant," problems described as "real world" are often introduced. The following example of such a problem is similar to many fourth grade assessment problems: *The picture below shows a 5 × 5 section of an array of lockers with only the 3 × 3 center group numbered.*

	11	12	13	
	20	21	22	
	29	30	31	

Figure 1

Students are given the following assessment task: *Some of the numbers have fallen off the doors of some old lockers. Figure out the missing numbers and describe the number pattern.*

This problem does not make sense mathematically. The data given are insufficient to find a unique answer. In fact, the expected "solution," as shown in figure 2, makes use of the *hidden assumption* that the array was rectangular. However, the assumptions that are given do not indicate that this is the case, and it would be improper, mathematically, to also assume that the array is rectangular.

1	2	3	4	5
10	11	12	13	14
19	20	21	22	23
28	29	30	31	32
37	38	39	40	41

Figure 2

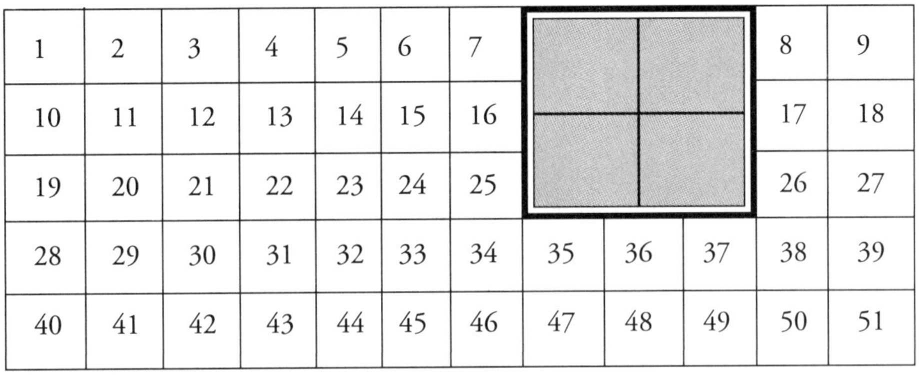

1	2	3	4	5	6	7			8	9
10	11	12	13	14	15	16			17	18
19	20	21	22	23	24	25			26	27
28	29	30	31	32	33	34	35	36	37	38	39
40	41	42	43	44	45	46	47	48	49	50	51

Figure 3

There are many other solutions without this assumption. For example, one is shown in figure 3.

One of the key points of mathematics is to promote critical thinking. Students have to learn to reason precisely with the data given so that if assumptions are hidden, they know they must seek them out and question them.

These remarks are not meant to diminish the importance of learning the number system and basic arithmetic, both of which are crucial as well. Here, too, these topics present problems for the kindergarten through grade seven curriculum, but not to the same degree as in many of the other areas discussed previously.

The intent of the material that follows in this chapter is to try to place into correct perspective much of the material taught in these grades, to indicate where problems might be encountered with some of the most important of these topics, and to suggest some ways of resolving the difficulties. In addition, throughout this chapter some items are pointed out to show where careful development will help foster critical thinking, and suggestions are given for accomplishing this process.

One of the key points of mathematics is to promote critical thinking.

Kindergarten Areas of Emphasis

By the end of kindergarten, students understand small numbers, quantities, and simple shapes in their everyday environment. They count, compare, describe, and sort objects and develop a sense of properties and patterns.

Number Sense

1.0 1.1 1.2 1.3

2.0 **2.1**

3.0 3.1

Algebra and Functions

1.0 **1.1**

Measurement and Geometry

1.0 1.1 1.2 1.3 1.4

2.0 2.1 2.2

Statistics, Data Analysis, and Probability

1.0 1.1 **1.2**

Mathematical Reasoning

1.0 1.1 1.2

2.0 2.1 2.2

Key Standards

NUMBER SENSE

The Number Sense standard that follows is basic in kindergarten:

 Students understand the relationship between numbers and quantities (i.e., that a set of objects has the same number of objects in different situations regardless of its position or arrangement).

A key skill within this standard is to group and compare sets of concrete items and recognize whether there are more, fewer, or an equal number of items in different sets. The following Number Sense standard is also important:

 Use concrete objects to determine the answers to addition and subtraction problems (for two numbers that are each less than 10).

The object of these standards is to begin to develop a precise sense of what a number is. Although students at this stage are dealing mainly with small numbers, they also need experience with larger numbers. An activity to provide this experience is to have the teacher fill glass jars with tennis balls, ping-pong balls, or jelly beans and ask the students to guess how many of these items are in the glass jar. Activities such as this one help give students an understanding of magnitude of numbers and help them gain experience with estimation.

When presenting this activity, teachers need to be aware that students can get the misconcept that larger numbers are only approximate rather than corresponding to exact quantities. This is a serious problem that has the potential to cause real difficulty later.

One way of avoiding this difficulty is to have the students use manipulatives, such as blocks, to compare two (relatively) large numbers; for example, 14 and 15. The class can explore the fact that 14 breaks up into two equal groups of 7, while 15 cannot be broken into two equal groups. The students would begin to appreciate that although visually distinguishing 15 objects from 14 without careful counting is difficult, the two numbers, nonetheless, are quite different. This activity should help students develop an awareness that each whole number is unique and will help them meet Number Sense Standard 1.2, which requires them to count and represent objects up to 30.

ALGEBRA AND FUNCTIONS

The role of the Algebra and Functions standard is also basic:

> **1.1** Identify, sort, and classify objects by attribute and identify objects that do not belong to a particular group (e.g., all these balls are green, those are red).

Although kindergarten teachers may not think of themselves as algebra teachers, they actually begin the process. They make students aware of the existence of patterns by giving them their first experience of finding them in data, by providing their initial exposure to functions, and by introducing them to abstraction. For example, students realize that a blue rectangular block and a blue ball, which obviously have different physical attributes, can nevertheless be sorted together because of their common color. This realization is the beginning of abstract reasoning, which is a higher-order thinking skill.

STATISTICS, DATA ANALYSIS, AND PROBABILITY

This standard interacts with the following Statistics, Data Analysis, and Probability standard:

> Identify, describe, and extend simple patterns (such as circles or triangles) by referring to their shapes, sizes, or colors.

Elaboration

The kindergarten teacher is likely to find that many students can learn more material than is specified in the kindergarten standards. For example, the standard for committing addition and subtraction facts to memory appears in the first grade. Because committing facts to memory requires substantial amounts of practice over an extended period, memorizing addition and subtraction facts can begin in kindergarten with simple facts, such as +1s, +2s, –1s, or sums to 10. Any practice of addition and subtraction facts should be limited to these more simple problems. Likewise, students can be taught the meaning of the symbols +, –, and = in the context of addition or subtraction, but again the focus is on small numbers. In measurement, the months can be taught in kindergarten as students learn the days of the week.

Considerations for Grade-Level Accomplishments in Kindergarten

Kindergarten is a critical time for children who, when they enter school, are behind their peers in the acquisition of skills and concepts. Efficient teaching in kindergarten can help prepare these children to work at an equal level with their peers in the later grades.

Students who enter kindergarten without some background in academic language (the language of tests and texts) and an understanding of the concepts such language represents have a great disadvantage in learning mathematics. Critical for beginning mathematical development are attributes, such as color, shape, and size; abstract concepts, such as *some, all,* and *none;* and ordinal concepts, such as *before, after, yesterday,* and *tomorrow.* Teachers need clear directions on how to maximize progress in mathematics for students with limited understanding of language concepts or for students who know the concepts in their native language but do not yet know the English words for them. Kindergarten provides many opportunities for teachers to teach basic mathematics vocabulary and concepts during instructional time or playtime; for example, students learn to take turns during a game or line up for recess (first, second, third), count off in a line (one, two, three), or determine the number of children who can take six balls out for recess if each child gets a ball (matching sets).

The most important mathematical skills and concepts for children in kindergarten to acquire are described as follows:

- **Counting.** Before beginning instruction in counting, teachers should determine the number to which the child can already count and whether the child understands what each number represents. The teacher models the next few numbers in the sequence (e.g., 5, 6, 7); provides practice for the children in saying the counting sequence through the new numbers (1, 2, 3, 4, 5, 6, 7); and matches each number to a corresponding set of objects. After a student has mastered the sequence including the new numbers, the teacher introduces several more numbers and follows the same procedure. Even though the standard requires a mastery of counting only to 30, daily practice in counting can be provided until students can count to 50 or 100 so that they may be better prepared for the challenges of the first grade.

- **Reading numerals.** The teacher should introduce numerals after the children can count to 10. Confusion between numeral names and the counting order can be *decreased* if the teacher does not introduce the numerals in order. For example, the teacher introduces the numeral 4 and then 7. For several days the teacher introduces a new numeral until the students can identify the numerals 1 through 10. The teacher should provide cumulative practice by having students review previously introduced numbers while he or she presents a new number.

- **Writing numerals.** The standards require that students know the names of the numerals from 1 to 9 and how to write them. Generally, writing numbers will require a good deal of practice; and at this age some children may have difficulty with coordination.

First, students should copy a numeral many times. Then they should write it with some prompts (e.g., dots or arrows); and later they should write it from memory, with the teacher saying the number and the student writing the numeral. A multisensory approach is very important here. Teachers must encourage the students of this age not to be concerned about the quality of their handwriting as they write numerals. Young children do not yet have fully developed fine-motor skills. Many students become frustrated by the discrepancy between what they want to produce on paper and what they can actually produce.

- **Understanding place value—reading numbers in the teens.** To read and write numbers from 10 to 20, students will need to understand something about place value. The teacher can expect the numbers 11, 12, 13, and 15 to be more troublesome than 14, 16, 17, 18, and 19. The second group is regular in pronunciation (e.g., *four*teen, *six*teen), but the first group is irregular; twelve is not pronounced as "twoteen" but as "twelve."

An important prerequisite for understanding place value is being able to answer fact questions verbally; for example, what is 10 + 6? When the students know the facts about numbers in the teens that are regular in pronunciation, the teacher can introduce one number with irregular pronunciation and mix it with the regular numbers in a verbal exercise. New irregular numbers can be introduced as students demonstrate knowledge of previously introduced facts about numbers in the teens. Reading and writing these numbers can be introduced when students are able to do the verbal exercises.

- **Learning the days of the week.** The days of the week can be taught in a manner similar to that for counting, in which the teacher models a part of the sequence of days (Monday, Tuesday, Wednesday); provides practice in saying the sequence; introduces a new part after several days (Thursday, Friday); provides practice with this part; and then repeats the sequence from the beginning. The months of the year can also be taught in kindergarten. Unless the students have a firm understanding of the sequence of days and months, they will have difficulty with items applying concepts of time, such as *before* and *after* as indicated in the second part of the following standard:

MEASUREMENT AND GEOMETRY

 Students understand the concept of time and units to measure it; they understand that objects have properties, such as length, weight, and capacity, and that comparisons may be made by referring to those properties.

An important prerequisite for understanding place value is being able to answer fact questions verbally.

Grade One — Areas of Emphasis

By the end of grade one, students understand and use the concept of ones and tens in the place value number system. Students add and subtract small numbers with ease. They measure with simple units and locate objects in space. They describe data and analyze and solve simple problems.

Number Sense

1.0 **1.1** **1.2** 1.3 1.4 1.5

2.0 **2.1** **2.2** **2.3** **2.4** **2.5** 2.6 2.7

3.0 3.1

Algebra and Functions

1.0 1.1 1.2 1.3

Measurement and Geometry

1.0 1.1 1.2

2.0 2.1 2.2 2.3 2.4

Statistics, Data Analysis, and Probability

1.0 1.1 1.2

2.0 **2.1**

Mathematical Reasoning

1.0 1.1 1.2

2.0 2.1 2.2

3.0

Key Standards

NUMBER SENSE

The following Number Sense standard is basic:

 Count, read, and write whole numbers to 100.

It is important that students gain a conceptual understanding of numbers and counting, not simply learn to count to 100 by rote. They need to understand, for example, that counting can occur in any order and in any direction, not just in the standard left-to-right counting pattern, as long as each item is tagged once and only once. Students must understand that numbers represent sets of specific quantities of items. Of particular importance is learning and understanding the counting sequence for numbers in the teens and multiples of ten. It should be emphasized that numbers in the teens represent a ten value and a certain number of unit values—12 does not merely represent a set of 12 items; it also represents 1 ten and 2 ones. A related and equally important Number Sense standard is:

- **1.2** Compare and order whole numbers to 100 by using the symbols for less than, equal to, or greater than (<, =, >).

The continuing development of addition and subtraction skills as described in the following standards is basic:

- **2.1** Know the addition facts (sums to 20) and the corresponding subtraction facts and commit them to memory.
- **2.5** Show the meaning of addition (putting together, increasing) and subtraction (taking away, comparing, finding the difference).

For example, students should understand that the equation $15 - 8 = 7$ is the same as $15 = 7 + 8$. Particular attention should be paid to the assessment of these competencies because students who fail to learn these topics will have serious difficulties in the later grades. The achievement of these standards will require that students be exposed to and asked to solve simple addition and subtraction problems throughout the school year.

STATISTICS, DATA ANALYSIS, AND PROBABILITY

The following Statistics, Data Analysis, and Probability standard is also important, but it has to be handled carefully:

- **2.1** Describe, extend, and explain ways to get to a next element in simple repeating patterns (e.g., rhythmic, numeric, color, and shape).

Students should *never* get the idea that the next term *automatically* repeats (unless they are told explicitly that it does); however, it is legitimate to ask what is the *most likely* next term. In this way students begin to learn not only the usefulness of patterns in sorting and understanding data but also careful, precise patterns of thought. Examples are sequences of colors, such as red, blue, red, blue, . . . or numbers, 1, 2, 3, 1, 2, 3, 1, 2, 3, . . . But more complex series might also be used, such as 1, 2, 3, 2, 1, 2, 3, 2, 1, 2, 3, . . .

Elaboration

Teaching students to solve basic addition and subtraction problems effectively and to commit the answers to memory will require considerable practice in solving these problems. As described in Chapter 4, the associated practice should be in small doses each day or, at the very least, several times a week. At the beginning of the school year, practice should focus on smaller problems (with sums less than or equal to ten). Large-valued problems should be emphasized in practice once students are skilled at solving the easier problems. Frequent assessment should be provided to determine whether students are mastering new facts and retaining those taught previously. Students have mastered basic facts when they can solve problems involving those facts quickly and accurately. Accurate but slow problem solving indicates that students are still using counting or other procedures to solve simple problems and have not yet committed the basic facts to memory.

Committing the basic addition and subtraction facts to memory is a major objective in the first and second grades. Students who do not commit the basic facts to memory will be at a disadvantage in further work with numbers and arithmetic.

Students have mastered basic facts when they can solve problems involving those facts quickly and accurately.

Grade One

The symmetric relationship between sets of simple addition problems, such as 7 + 2 and 2 + 7, can be used to reduce the memorization load in learning facts.

Understanding the symmetric relationship between sets of simple addition problems, such as 7 + 2 and 2 + 7, can be used to reduce the memorization load in learning facts. The teaching of these relationships is to be incorporated into the sequence for teaching students simple addition and during their practice. For example, after students have learned 7 + 2, they can be shown that the same answer applies to 2 + 7. Moreover, by placing problems such as 7 + 2 and 2 + 7 in sequence in practice sheets, students will have the opportunity to "discover" and reinforce this relationship as well. Later, they might learn that the combination of 7, 2, and 9 can be used to create subtraction facts and addition facts.

While the standard calls for counting by 1 to 100 in the first grade, counting into the 100s can begin in the latter part of the first grade if students have mastered counting to 100. Counting backward for numbers up to 100 should also be done in the first grade once students have mastered counting forward.

Considerations for Grade-Level Accomplishments in Grade One

The most important mathematical skills and concepts for children in grade one to acquire are described as follows:

- **Reading and writing of numbers.** Many students demonstrate a lack of understanding of place value when they encounter numbers such as 16 and 61. If students are confused by two such similar numbers, teachers should try to determine whether the cause of the confusion is students' failure to understand that numbers are read from left to right or students' inadequate understanding of place value. Instruction should be carefully sequenced to show that 16 is 1 ten and 6 ones, while 61 is 6 tens and 1 one. Students need to know prerequisite skills underlying place value, such as 6 tens equals 60 and its corollary, 60 equals 6 tens, and addition facts in which a single-digit number is added to the tens number, 10 + 3, 10 + 5, 30 + 6. These facts can be taught verbally before students read and write the numbers.

  Learning the number that represents a group of tens is important for understanding place value and reading numbers. Some students are more likely to have difficulty with groups of tens in which the tens number does not say the name of the first digit (e.g., "twenty" is not pronounced "twoty") than with tens numbers in which the name of the first digit is pronounced, sixty, forty, seventy, eighty, ninety. Teachers should provide more practice on the more difficult items.

- **Skip counting.** In addition to enhancing children's number sense, skip counting is important for facilitating the learning of multiplication and division. Counting by tens should be introduced when students can count by ones to about 20 or 30. Counting by tens helps students learn to count by ones to 100. Skip counting is taught just like counting by ones. The teacher models the first part of the sequence; then the students practice the first part. The modeling and practicing continue on new parts of the sequence until students can say the whole sequence. Skip counting requires systematic teaching using a procedure similar to that discussed for counting by ones. Regularly scheduled practice will help students master counting a sequence. Previously introduced sequences should be reviewed as students learn new ones.

- **Teaching of addition and subtraction facts.** Teaching addition and subtraction facts and making assessments should be systematic, as was discussed previously.
- **Understanding of symmetric relationships.** Understanding the symmetric relationship of facts can reduce the number of facts to be memorized in learning.
- **Adding and subtracting of one- and two-digit numbers.** Students can be helped to avoid difficulties with adding one- and two-digit numbers if they are given practice with "lining up" numbers in the problem and adding from right to left. This procedure can be confusing to students because (as previously discussed) we read and write numbers from left to right. Furthermore, in anticipation of subtracting one- and two-digit numbers, students need practice in working from top to bottom.

Grade Two — Areas of Emphasis

By the end of grade two, students understand place value and number relationships in addition and subtraction, and they use simple concepts of multiplication. They measure quantities with appropriate units. They classify shapes and see relationships among them by paying attention to their geometric attributes. They collect and analyze data and verify the answers.

Number Sense

1.0 **1.1** 1.2 **1.3**

2.0 **2.1** **2.2** 2.3

3.0 **3.1** **3.2** **3.3**

4.0 **4.1** **4.2** **4.3**

5.0 **5.1** **5.2**

6.0 6.1

Algebra and Functions

1.0 **1.1** 1.2 1.3

Measurement and Geometry

1.0 1.1 1.2 **1.3** 1.4 1.5

2.0 **2.1** **2.2**

Statistics, Data Analysis, and Probability

1.0 1.1 1.2 1.3 1.4

2.0 2.1 2.2

Mathematical Reasoning

1.0 1.1 1.2

2.0 2.1 2.2

3.0

Key Standards

NUMBER SENSE

As was the case in grade one, the students' growing mastery of whole numbers is the basic topic in grade two, although fractions and decimals now appear. These Number Sense standards are particularly important:

- **1.1** Count, read, and write whole numbers to 1,000 and identify the place value for each digit.
- **1.3** Order and compare whole numbers to 1,000 by using the symbols <, =, >.

Here, the reiteration of the use of the symbols <, =, > serves the additional role of giving the students an example of abstraction and the key algebraic process of substituting symbols for more complex concepts and quantities. Because the symbols < and > are confusing to many children, they need to use them frequently.

For many of the same reasons, the standards listed below are very important:

- **2.1** Understand and use the inverse relationship between addition and subtraction (e.g., an opposite number sentence for $8 + 6 = 14$ is $14 - 6 = 8$) to solve problems and check solutions.
- **2.2** Find the sum or difference of two whole numbers up to three digits long.

Standard 2.1 gives students a clear application of the relations between different types of operations (addition and subtraction) and can be used to encourage more flexible methods of thinking about and solving problems; for example, a knowledge of addition can facilitate the solving of subtraction problems and vice versa. The problem $144 - 98 = ?$ can be solved by realizing that $144 = 100 + 44 = 98 + 2 + 44 = 98 + 46$.

The third Number Sense standard is basic to students' understanding of arithmetic and the ability to solve multiplication and division problems:

- **3.0** Students model and solve simple problems involving multiplication and division.

Here, fluency with skip counting is helpful.

The discussion of fractions and the goals represented in Number Sense Standards 4.1, 4.2, and 4.3 are also essential features of students' developing arithmetical competencies. Although equivalence of fractions is not explicitly presented in the standards, it is also a good idea to begin the discussion of the topic at this point—students should know, for example, that $2/4$ is the same as $1/2$, a concept that can (and should) be demonstrated with pictures. Finally, as a practical matter and as a basic application of the topics discussed previously, the material in Number Sense Standards 5.1 and 5.2—on modeling and solving problems involving money—is very important. Borrowing money gives a practical context to the concept of subtraction. Special attention should be paid to the need for introducing the symbols $ and ¢ and to the fact that the order of the symbol for dollars is $3, not 3$; but for cents, the order is 31¢, not ¢31.

ALGEBRA AND FUNCTIONS

In the Algebra and Functions strand, the following standard is an *essential* feature of mathematics instruction in grade two:

- **1.1** Use the commutative and associative rules to simplify mental calculations and to check results.

However, the emphasis here should be on the *use* of these rules to simplify; for example, knowing that $5 + 8 = 13$ saves the labor of also learning that $8 + 5 = 13$.

> Although equivalence of fractions is not explicitly presented in the standards, it is a good idea to begin the discussion of the topic at this point.

Learning the terminology is not nearly as important. The students should begin to develop an appreciation for the power of unifying rules; but *overemphasizing these topics, particularly the sophisticated concept of the associative rule, is probably worse than not mentioning them at all.*

MEASUREMENT AND GEOMETRY

Although Standard 1.3 listed below from the Measurement and Geometry strand is important, more emphasis should be given to the topics in Standard 2.0.

- **1.3** Measure the length of an object to the nearest inch and/or centimeter.
- **2.0** Students identify and describe the attributes of common figures in the plane and of common objects in space.

Because understanding spatial relations will be more difficult for some students than for others (especially the concepts involving three-dimensional information), teachers should carefully assess how well students understand these shapes and figures and their relationships.

STATISTICS, DATA ANALYSIS, AND PROBABILITY

Although Standard 1.0 in the Statistics, Data Analysis, and Probability strand is important for grade two, the topics in Standard 2.0 are more important in this grade.

- **1.0** Students collect numerical data and record, organize, display, and interpret the data on bar graphs and other representations.
- **2.0** Students demonstrate an understanding of patterns and how patterns grow and describe them in general ways.

But here, as for grade one, it is important that students distinguish between the most likely next term and *the* next term. In statistics students look for likely patterns, but in mathematics students need to know the rule that generates the pattern to determine "the" next term. As an example, given only the sequence 2, 4, 6, 8, 10, students should *not* assert that the next term is 12 but, instead, that the most likely next term is 12. For example, the series might have actually been 2, 4, 6, 8, 10, 14, 16, 18, 20, 22, 26, 28 . . . . The ability to distinguish between what is likely and what is given promotes careful, precise thought.

Elaboration

In the second grade, work on committing the answers to basic addition and subtraction problems to memory should continue for those students who have not mastered them in the first grade. Students' knowledge of facts needs to be assessed at the beginning of the school year. The assessment could be done individually so that the teacher can determine whether the student has committed the facts to memory. Mastery of addition and subtraction facts can also be assessed with simple paper-and-pencil tests. Students should be asked to solve a whole sheet of problems in one or two minutes. As noted earlier, students who have committed the basic facts to memory will quickly and correctly dispose of these simple tasks. If not, they are, most likely, solving the problem by counting in their head (Geary 1994) or using time-consuming counting procedures to generate answers. Additional practice will be necessary for these children.

Students learn the basics of how to "carry" and "borrow" in the second grade. Because carrying and borrowing are difficult for students to master, extended discussion and practice of these skills will likely be necessary (Fuson and Kwon 1992). To carry and borrow correctly, it is important that students understand the base-10 structure of the number system and the concept that carrying and borrowing involve exchanging sets of 10 ones or 10 tens and so forth from one column to the next. It is common for students to incorrectly conceptualize carrying or borrowing; for example, taking a one from the tens column and giving it to the ones column. What has been given, in fact, is one set of 10 units, not one unit from the tens. For example, borrowing in the case of 43 − 7 can be explained as follows: 43 − 7 = (30 + 13) − 7 = 30 + (13 − 7) = 30 + 6 = 36, illustrating the associative law of addition in the process. Initially, problems should be limited to those that require carrying or borrowing across one column (e.g., 17 + 24, 43 − 7), and particular attention should be paid to problems with zero (90 − 34 and 94 − 30) because they are often confusing to students (VanLehn 1990).

Multiplication is introduced in the second grade, and students are to commit to memory the twos, fives, and tens facts. During the initial learning of multiplication, students often confuse addition and multiplication facts, but these confusions should diminish with additional practice. These facts should be taught with the same systematic approach as was discussed for the addition facts in grade one. *The skip counting series for numbers other than 2, 5, and 10 (e.g., 3s, 4s, 9s, 7s, 25s) can be introduced in the second grade to prepare students for learning more multiplication facts in the third grade.* Additionally, the associative and commutative laws can be used to increase the number of multiplication facts the students know. For example, there is no need for students to learn 5 × 8 if they already know 8 × 5.

Students in these early grades often have trouble lining numbers up for addition or subtraction. Reminding students to make sure that their numbers are lined up evenly is essential. Students can be taught to use estimation to determine whether their answers are reasonable. However, it is unwise to try to put undue emphasis on estimation by teaching second grade students to answer problems only by making estimates. Instead, they should concentrate on problems that demand an exact answer and use estimation to check whether their answer is reasonable.

The work with fractions should include examples showing fractions that are less than one, fractions that are equal to one, and fractions that are equal to more than one. This range is needed to prevent students from thinking that fractions express only units less than one. To this end, teachers need to make sure that students can freely work with improper fractions and understand that, the name notwithstanding, there is nothing wrong with improper fractions.

It has been pointed out that many second grade students have real difficulty with the written form of fractions but much less trouble with their verbal descriptions. Therefore, the verbal descriptions should be emphasized at this level, although students will, of course, eventually need to know the standard written representations of fractions.

Considerations for Grade-Level Accomplishments in Grade Two

The most important mathematical skills and concepts for children in grade two to acquire are described as follows:

- **Counting.** Many students require careful teaching of counting from 100 through 999. Students can learn the counting skills for the entire range through exercises in which the teacher models and provides practice sets consisting of series. First, the teacher

Borrowing illustrates the associative law of addition.

models numbers within a particular decade (e.g., 350, 351, 352, 353, 354, 355, 356, 357, 358, 359, 360). A daily teaching session might include work on several series (e.g., 350 to 360, 140 to 150, 470 to 480). Sets within a decade would be worked on daily until students demonstrate the ability to generalize to new series. During the next stage students would practice on series in which they move from one decade to the next (e.g., 365 to 375, 125 to 135, 715 to 725). Students may have difficulty making the transition from one decade to the next without explicit instruction and adequate practice. When the students demonstrate a general ability to make this transition, the final set of series would be introduced. These sets would include those in which the transition from one one-hundred number to the next occurs: 595 to 605, 195 to 205, 495 to 505.

- **Writing numbers.** If the students are not instructed carefully, some may develop the misconception that the presence of two zeros creates a hundreds number. These students will write three hundred twenty-five as 30025. Teachers should watch for this type of error and correct it immediately. Examples with and without zeros need to be modeled and practiced.

- **Borrowing.** Practice with the terms *more* and *less* and *top* and *bottom* should precede the introduction of problems involving borrowing. These concepts need to be firmly understood if students are to succeed with borrowing problems.

- **Skip counting.** In the process of using skip counting to learn multiples, students may become confused by numbers that appear on several lists. For example, when numbers are counted by threes and fours, the number 12 appears as the fourth number on the "multiples of three" list and as the third number on the "multiples of four" list. To avoid confusing their students, teachers should provide extensive practice with one of these sequences before introducing the next.

- **Counting groups of coins.** This process requires that students be able to say the respective count by series for the value of each coin and be able to answer addition fact questions easily, such as 25 + 5, 30 + 10, in which a nickel or dime is added to a number ending in 5 or 0. Exercises in counting coins should be coordinated with instruction in counting facts so that students have already practiced the skill thoroughly before having to apply it. Counting coins should be reviewed and extended to include quarters along with dimes, nickels, and pennies. A particular fact that some students find difficult to comprehend is adding ten to a two-digit number ending in 5 (e.g., 35 + 10).

- **Aligning columns.** Students may need systematic instruction in rewriting problems written as a column problem; practice in rewriting horizontal equations, such as 304 + 23 =__ or 6 + 345 = ___, in column form; and help in lining numbers up for addition or subtraction. In certain situations they can be taught to use estimation to check whether their answers are reasonable and, if not, to recheck their work to find their mistakes. As was discussed previously in the subsection on elaboration, it is unwise to try to teach students in grade two to answer problems that request only an estimate as the answer. Students need to become accustomed to obtaining exact answers and using estimation only as an aid to check whether the answer is reasonable.

- **Reviewing time equivalencies.** Students will need to review time equivalencies (e.g., 1 minute equals 60 seconds, 1 hour equals 60 minutes, 1 day equals 24 hours, 1 week equals 7 days, 1 year equals 12 months). These equivalencies need to be practiced and reviewed so that all students are able to commit them to memory.
- **Understanding money.** In the teaching of decimal notation for money, teachers must ensure that students can read and write amounts such as $2.05, in which there is a zero in the tenths column, and $.65, in which there is no dollar amount. By the end of the second grade, students should be able to write ten cents as $.10 and ten dollars as $10.00 in decimal notation.
- **Telling time.** Students can be taught a general procedure for telling time. Telling time on an analog clock can begin with teaching students to tell how many minutes after the hour, to the nearest five minutes, are shown on the clock. Students need to be proficient in counting by fives before time telling is introduced. When the students can read the minutes after the hour, reading the minutes before the hour can be introduced. Students should be taught to express the time as minutes after and as minutes before the hour (e.g., 40 minutes after 1 is the same as 20 minutes before 2).
- **Understanding fractions.** Creating a fraction to represent the parts of a whole (e.g., $2/3$ of a pie) is significantly different from dividing a set of items into subgroups and determining the number of items within some subgroups (e.g., $2/3$ of 15). A unit divided into parts can be introduced first, and instruction on that type of fraction should be provided until students can recognize and write fractions to represent fractions of a whole; then the more complex fractions should be introduced. Students can work with diagrams. Computer programs and videos are also available to help with this topic. Students are not expected to solve $2/3$ of 15 numerically in the second grade, because doing so requires them to be able to multiply fractions and convert an improper fraction to a whole number.

> **Grade Two**
>
> Students need to become accustomed to obtaining exact answers and using estimation only as an aid to check the answer.

Grade Three | Areas of Emphasis

By the end of grade three, students deepen their understanding of place value and their understanding of and skill with addition, subtraction, multiplication, and division of whole numbers. Students estimate, measure, and describe objects in space. They use patterns to help solve problems. They represent number relationships and conduct simple probability experiments.

Number Sense

1.0 1.1 1.2 **1.3** 1.4 **1.5**
2.0 **2.1** **2.2** **2.3** **2.4** 2.5 2.6 2.7 2.8
3.0 3.1 **3.2** **3.3** 3.4

Algebra and Functions

1.0 **1.1** 1.2 1.3 1.4 1.5
2.0 **2.1** 2.2

Measurement and Geometry

1.0 1.1 **1.2** **1.3** 1.4
2.0 **2.1** **2.2** **2.3** 2.4 2.5 2.6

Statistics, Data Analysis, and Probability

1.0 1.1 **1.2** **1.3** 1.4

Mathematical Reasoning

1.0 1.1 1.2
2.0 2.1 2.2 2.3 2.4 2.5 2.6
3.0 3.1 3.2 3.3

Key Standards

NUMBER SENSE

In the Number Sense strand, Standards 1.3 and 1.5 are especially important:

1.3 Identify the place value for each digit in numbers to 10,000.

1.5 Use expanded notation to represent numbers (e.g., 3,206 = 3,000 + 200 + 6).

For students who show a good conceptual understanding of whole numbers (e.g., place value), the second standard should receive special attention. Here, Standards 2.1, 2.2, 2.3, and 2.4 are especially important:

- **2.1** Find the sum or difference of two whole numbers between 0 and 10,000.
- **2.2** Memorize to automaticity the multiplication table for numbers between 1 and 10.
- **2.3** Use the inverse relationship of multiplication and division to compute and check results.
- **2.4** Solve simple problems involving multiplication of multidigit numbers by one-digit numbers ($3,671 \times 3 =$ ___).

Two topics in the third standard also deserve special attention:

- **3.2** Add and subtract simple fractions (e.g., determine that $\frac{1}{8} + \frac{3}{8}$ is the same as $\frac{1}{2}$).
- **3.3** Solve problems involving addition, subtraction, multiplication, and division of money amounts in decimal notation and multiply and divide money amounts in decimal notation by using whole-number multipliers and divisors.

These are the early introductory elements of arithmetic with fractions and decimals—topics that will build over several years.

ALGEBRA AND FUNCTIONS

In the third grade, the Algebra and Functions strand grows in importance:

- **1.1** Represent relationships of quantities in the form of mathematical expressions, equations, or inequalities.

Because understanding these concepts can be a very difficult step for students, instruction must be presented carefully, and many examples should be given: 3×12 inches in 3 feet, 4×11 legs in 11 cats, 2×15 wheels in 15 bicycles, 3×15 wheels in 15 tricycles, the number of students in the classroom < 50, the number of days in a year > 300, and so forth.

The next three standards expand on the first and provide examples of what is meant by "represent relationships of . . . ." Teachers must be sure that students are aware of the power of commutativity and associativity in multiplication as a simplifying mechanism and as a means of avoiding overemphasis on pure memorization of the formulas without understanding.

The second standard is also important and likewise must be treated carefully:

- **2.1** Solve simple problems involving a functional relationship between two quantities (e.g., find the total cost of multiple items given the cost per unit).

MEASUREMENT AND GEOMETRY

In the first Measurement and Geometry standard, Standards 1.2 and 1.3 should be emphasized:

- **1.2** Estimate or determine the area and volume of solid figures by covering them with squares or by counting the number of cubes that would fill them.
- **1.3** Find the perimeter of a polygon with integer sides.

The idea that one cannot talk about area until a square of side 1 has been declared to have unit area and is then used to measure everything else is usually not firmly established in standard textbooks. Analogies should be constantly drawn between length and area. For example, a line segment having a length 3 means that, compared with the segment L that has been declared to be of length 1, it can be covered exactly by 3 nonoverlapping copies of *L*. Likewise, a rectangle with sides of lengths 3 and 1 has an area equal to 3 because it can be exactly covered by three nonoverlapping copies of the square declared to have length 1.

In the second Measurement and Geometry standard, Standards 2.1, 2.2, and 2.3 are the most important.

2.1 Identify, describe, and classify polygons (including pentagons, hexagons, and octagons).

2.2 Identify attributes of triangles (e.g., two equal sides for the isosceles triangle, three equal sides for the equilateral triangle, right angle for the right triangle).

2.3 Identify attributes of quadrilaterals (e.g., parallel sides for the parallelogram, right angles for the rectangle, equal sides and right angles for the square).

All of these standards can be difficult to master if they are presented too generally. It is strongly recommended that the skills for this grade level be limited to such topics as finding the areas of rectangles with integer sides, right triangles with integer sides, and figures that can be partitioned into such rectangles and right triangles. A few examples in which the sides are not whole numbers should also be provided. Estimation should be used for these examples. Implicit in Standards 2.4 and 2.5 is the introduction of the concept of an angle. But this topic should not be emphasized at this time.

STATISTICS, DATA ANALYSIS, AND PROBABILITY

The most important standards for Statistics, Data Analysis, and Probability are:

1.2 Record the possible outcomes for a simple event (e.g., tossing a coin) and systematically keep track of the outcomes when the event is repeated many times.

1.3 Summarize and display the results of probability experiments in a clear and organized way (e.g., use a bar graph or a line plot).

Elaboration

In the third grade, work with addition and subtraction problems expands to problems in which regrouping (i.e., carrying and borrowing) is required in more than one column. As noted earlier particularly important and difficult for some students are subtraction problems that include zeros; for example, $302 - 25$ and $3002 - 75$ (VanLehn 1990). Students need to become skilled in regrouping across columns with zeros because such problems are often used with money applications; for example, *Jerry bought an ice cream for 62 cents and paid for it with a ten-dollar bill. How much change will he receive?*

One way to treat $302 - 25$ is again through the use of the associative law of addition: $302 - 25 = (200 + 102) - 25 = 200 + (102 - 25) = 200 + (2 + 100 - 25) = 200 + (2 + 75) = 277$. The first equality is exactly what is meant by "borrowing in the 100s place."

As with addition and subtraction, memorizing the answers to simple multiplication problems requires the systematic introduction and practice of facts. (Refer to the recommendations discussed for addition facts in the first-grade section.) Some division facts can be incorporated into the sequence for learning multiplication facts. As with addition and subtraction, symmetric relationships can be used to cut down on the need for memorization. These related facts can be introduced together (20 divided by 5, 5 times 4).

Multiplication and division problems with multidigit terms are introduced in the third grade (e.g., 36 × 5). The basic facts used in both types of problems should have already been committed to memory (e.g., students should have already memorized the answer to 6 × 5, a component of the more complex problem 36 × 5). Students should already be familiar with the basic structure of these problems because of their understanding of how to add a one-digit to a two-digit number (e.g., 18 + 4 and 36 + 5, 12 + 6). As with addition and subtraction, problems that require carrying (e.g., 36 × 5) will be more difficult to solve than will the problems that do not require carrying (e.g., 32 × 4) (Geary 1994).

The goal is to extend the multiplication of whole numbers up to 10,000 by single-digit numbers (e.g., 9,345 × 2) so that students gain mastery of the standard right-to-left multiplication algorithm with the multiplier being a one-digit number.

Students are expected to work on long division problems in which they divide a multidigit number by a single digit. A critical component skill for solving these problems is the ability to determine the multiple of the divisor that is just smaller than the number being divided. In $^{28}/_5$, the multiple of 5 that is just smaller than 28 is 25. Although the identification of remainders exceeds the level of the third grade standard, students need to become aware of the process for division when there is a remainder. Practice in determining multiples can be coordinated with the practice of multiplication facts. Having basic multiplication facts memorized will greatly facilitate students' ability to solve these division problems.

Rounding is a critical prerequisite for working estimation problems. Noted in the next column is a sequence of exercises that might be followed when introducing rounding. Each exercise can be introduced over several days, followed by continued practice. Practice sets should include examples that review earlier stages and present the current ones, as described in Appendix A, "Sample Instructional Profile."

- Round a 2-digit number to the nearest 10.
- Round a 3-digit number to the nearest 10.
- Round a 3-digit number to the nearest 100.
- Round a 4-digit number to the nearest 1,000.
- Round a 4-digit number to the nearest 100.

The work with fractions in grade three is primarily with diagrams and concrete objects. Students should be able to compare fractions in at least two ways. First, students should be able to order fractions—proper or improper—with like denominators, initially using diagrams but later realizing that if the denominators are equal, then the order depends only on the numerators. Second, students should be able to order unit fractions, perhaps only with whole-number denominators less than or equal to 6. It is not expected at this point that students should compare fractions with unlike denominators except for very simple cases, such as ¼ and ⅜ or ½ and ¾. Students should compare particular fractions verbally and with the symbols <, =, >.

With regard to multiplying and dividing decimals, care should be taken to include exercises in which students have to distinguish between adding and multiplying. Work with money can serve as an introduction to decimals. For example, the following problem is typical of the types

Grade Three

Rounding is a critical prerequisite for working estimation problems.

of problems that can serve as the introduction of decimal addition:

Josh had $3. He earned $2.50. How much does he have now?

Likewise, the next problem typifies the types of problems that can introduce decimal multiplication:

Josh earned $2.50 an hour. He worked 3 hours. How much did he earn?

The teaching of arithmetic facts can be extended in the third grade to include finding multiples and factors of whole numbers; both are critical to students' understanding of numbers and later to simplifying fractions. Because students need time to develop this skill, it is recommended that they be given considerable instruction on it before they are tested. Only small numbers involving few primes should be used. As a rule, "small" means less than 30, with prime factors limited to only 2, 3, or 5 (e.g., 20 = 2 × 2 × 5, 18 = 3 × 3 × 2).

Considerations for Grade-Level Accomplishments in Grade Three

The most important mathematical skills and concepts for children in grade three to acquire are described as follows:

- **Addition and subtraction facts.** Students who enter the third grade without addition and subtraction facts committed to memory are at risk of having difficulty as more complex mathematics is taught. An assessment of students' knowledge of basic facts needs to be undertaken at the beginning of the school year. Systematic daily practice with addition and subtraction facts needs to be provided for students who have not yet learned them.

- **Reading and writing of numbers.** Thousands numbers with zeros in the hundreds or tens place or both (4006, 4060, 4600) can be particularly troublesome for at-risk students. Systematic presentations focusing on reading and writing thousands numbers with one or two zeros need to be provided until students can read and write these more difficult numbers.

- **Rounding off.** Rounding off a thousands number to the nearest ten, hundred, and thousand requires a sophisticated understanding of the rounding-off process. When rounding to a particular unit, students need to learn at which point to start the rounding process. For example, when rounding off to the nearest hundred, the student needs to look at the current digit in the tens column to determine whether the digit in the hundreds column will remain the same or be increased when rounded off. Practice items need to include a variety of types (e.g., round off 2,375 to the nearest hundred and then to the nearest thousand).

- **Geometry.** While many of these geometric concepts are not difficult in themselves, students typically have difficulty, becoming confused as new concepts and terms are introduced. This problem is solvable through the use of a cumulative manner of introduction in which previously introduced concepts are reviewed as new concepts are introduced.

- **Measurement.** The standards call for students to learn a significant number of measurement equivalencies. These equivalencies should be introduced so that students are not overwhelmed with too much information at one time. Adequate practice and review are to be provided so that students can readily recall all equivalencies.

Grade Four | Areas of Emphasis

By the end of grade four, students understand large numbers and addition, subtraction, multiplication, and division of whole numbers. They describe and compare simple fractions and decimals. They understand the properties of, and the relationships between, plane geometric figures. They collect, represent, and analyze data to answer questions.

Number Sense

1.0 **1.1** **1.2** **1.3** **1.4** 1.5 1.6 1.7 **1.8** **1.9**

2.0 2.1 2.2

3.0 **3.1** **3.2** **3.3** **3.4**

4.0 4.1 **4.2**

Algebra and Functions

1.0 1.1 **1.2** **1.3** 1.4 **1.5**

2.0 **2.1** **2.2**

Measurement and Geometry

1.0 1.1 1.2 1.3 1.4

2.0 **2.1** **2.2** **2.3**

3.0 3.1 3.2 3.3 3.4 3.5 3.6 3.7 3.8

Statistics, Data Analysis, and Probability

1.0 1.1 1.2 1.3

2.0 2.1 2.2

Mathematical Reasoning

1.0 1.1 1.2

2.0 2.1 2.2 2.3 2.4 2.5 2.6

3.0 3.1 3.2 3.3

Key Standards

NUMBER SENSE

The Number Sense strand for the fourth grade continues the development of whole numbers to the millions and then moves on to ordering and comparing numbers to two decimal places. At this grade level there are two requirements for rounding, both of which must be presented carefully although the topics are not yet centrally important. Standards 1.5 and 1.7 relate to equivalencies between writing numbers in fractional and decimal notations. These standards require an understanding of fractions not only as parts of a whole but as the quotient of the numerator by the denominator (Standard 1.5). They will become important in the later grades, but they should not be overemphasized at this point.

The next standards are basic and new standards:

- **1.8** Use concepts of negative numbers (e.g., on a number line, in counting, in temperature, in "owing").
- **1.9** Identify on the number line the relative position of positive fractions, positive mixed numbers, and positive decimals to two decimal places.

These standards can be difficult for students to learn if the required background material—ordering of whole numbers and comparison of fractions and decimals—is not presented carefully. The importance of these standards requires that close attention be paid to assessment.

The third topic in the Number Sense strand is also especially important. This and its four substandards all involve the use of the standard algorithms for addition, subtraction, and multiplication of multidigit numbers as well as the standard algorithm for division of a multidigit number by a one-digit number. As with simple arithmetic, mastery of these skills will require extensive practice over several grade levels, as described in Chapter 4, "Instructional Strategies."

The fourth topic, "students know how to factor small whole numbers," is needed for the discussion of the equivalence of fractions. It also includes the requirement that students understand what a prime number is. The concept of primality is important yet often difficult for students to understand fully. Students should also know the prime numbers up to 50. For these reasons the preparation for the discussion of prime numbers should begin no later than the third grade. Students who understand prime numbers will find it easier to understand the equivalence of fractions and to multiply and divide fractions in grades five, six, and seven.

ALGEBRA AND FUNCTIONS

In the fourth grade the Algebra and Functions strand continues to grow in importance. All five of the subtopics under the first standard are important. But the degree to which students need to understand these strands differs. The following standards *do not need undue emphasis*:

- **1.2** Interpret and evaluate mathematical expressions that now use parentheses.
- **1.3** Use parentheses to indicate which operation to perform first when writing expressions containing more than two terms and different operations.

These standards involve nothing more than notation. The real skill is learning how to write expressions unambiguously so that others can understand them. However, it would be appropriate at this point to explain carefully to students why the associative and commutative laws are significant and why arbitrary sums or products, such as 115 + 6 + (−6) + 4792 or 113 × 212 × 31 × 11, do not have to be ordered in any particular way, nor do they have to be calculated in any particular order.

Standards 1.4 and 1.5, which relate to functional relationships, are much more important theoretically. In particular, students should understand Standard 1.5 because it takes the mystery out of the topic.

> **1.5** Understand that an equation such as $y = 3x + 5$ is a prescription for determining a second number when a first number is given.

One way to understand an equation such as $y = 3x + 5$ is to work through many pairs of numbers (x, y) to see if they satisfy this equation. For example, (1, 8) and (0, 5) do, but (−1, 3) and (2, 10) do not.

The second algebra standard is, however, basic:

> **2.0** Students know how to manipulate equations.

This standard and the two basic rules that follow, if understood now, will clarify much of what happens in mathematics and other subjects from the fifth grade through high school.

> **2.1** Know and understand that equals added to equals are equal.
>
> **2.2** Know and understand that equals multiplied by equals are equal.

However, if these concepts are not clear, difficulties in later grades are virtually guaranteed. Therefore, careful assessment of students' understanding of these principles should be done here.

MEASUREMENT AND GEOMETRY

The Measurement and Geometry strand for the fourth grade contains a few key standards that students will need to understand completely. The first standard (1.0) relates to perimeter and area. The students need to understand that the area of a rectangle is obtained by multiplying length by width and that the perimeter is given by a linear measurement. The intent of most of this standard is that students know the reasons behind the formulas for the perimeter and area of a rectangle and that they can see how these formulas work when the perimeter and area vary as the rectangles vary.

A more basic standard is the second one:

> **2.0** Students use two-dimensional coordinate grids to represent points and graph lines and simple figures.

Although the material in this standard is basic and is not presented in depth, this concept must be presented carefully. Again, students who are confused at this point will very likely have serious difficulties in the later grades—not just in mathematics, but in the sciences and other areas as well. Therefore, careful assessment is necessary.

In connection with Standard 3.0, teachers should introduce the symbol ⊥ for perpendicularity. Incidentally, this is the time to introduce the abbreviated notation *ab* in place of the cumbersome $a \times b$.

Elaboration

Knowledge of multiplication and division facts should be reassessed at the beginning of the school year, and systematic instruction and practice should be provided to enable students to reach high degrees of automaticity in recalling these facts. This process is described for addition in grade two (see "Elaboration").

Reading and writing thousands and millions numbers with one or more zeros in the middle can be particularly troublesome for students (Seron and Fayol 1994). Therefore, assessment and teaching should be thorough so that students are able to read and write difficult numbers, such as 300,200 and 320,000. Students need to understand that zeros in different positions represent different place values—tens, hundreds, thousands, and so forth—and they need practice in working with these types of numbers (e.g., determining which is larger, 320,000 or 300,200, and translating a verbal label, "one million two hundred thousand," into the Arabic representation, 1,200,000).

To be able to apply mathematics in the real world, to understand the way in which numbers distribute on the number line, and ultimately to study more advanced topics in mathematics, students need to understand the concept of "closeness" for numbers. It is probably not wise to push too hard on the notion of "close enough" while students are still struggling with the abstract idea of a number itself. However, by now they should be ready for this next step. A discussion of rounding should emphasize that one rounds off only if the result of rounding is "close enough."

Students need to understand fraction equivalencies related to the ordering and comparison of decimals. Students must understand, for instance, that 2/10 = 20/100, then equate those fractions to decimals.

The teaching of the conversion of proper and improper fractions to decimals should be structured so that students see relationships (e.g., the fraction 7/4 can be converted to 4/4 + 3/4, which in turn equals 1 and 3/4). The fourth grade standards do not require any arithmetic with fractions; however, practice with addition and subtraction of fractions (converting to like denominators) must be continued in this grade because these concepts are important in the fifth grade. Students can also be introduced to the concept of unlike denominators in preparation for the following year. Building students' skills in finding equivalent fractions is also important at this grade level.

The standards require that students know the definition of prime numbers and know that many whole numbers decompose into products of smaller numbers in different ways. Using the number 150 as an example, they should realize that $150 = 5 \times 30$ and $30 = 5 \times 6$; therefore, $150 = 5 \times 5 \times 6$, which can be decomposed to $5 \times 5 \times 3 \times 2$. Students will be using these factoring skills extensively in the later grades. Even though determining the prime factors of all numbers through 50 is a fifth grade standard, practice on finding prime factors can begin in the fourth grade. Students should be given extensive practice over an extended period of time with finding prime factors so that they can develop automaticity in the factoring process (see Chapter 4, "Instructional Strategies"). By the end of the fifth grade, students should be able to determine with relative ease whether any of the prime numbers 2, 3, 5, 7, or 11 are factors of a number less than 200.

Multiplication and division problems with multidigit numbers are expanded. Division problems with a zero in the quotient (e.g., 4233/6 = 705.5) can be particularly difficult for students to understand and require systematic instruction.

The Number Sense Standards 3.1 and 3.2 call for "understanding of the standard algorithm" (see the glossary). To present this concept, the teacher sketches the reasons why the algorithm works and carefully shows the students how to use it. (Any such explanation of the multiplication and division algorithms would help students to deepen their understanding and appreciation of the distributive law.) The students are not expected to reproduce this discussion in any detail, but they are expected to have a general idea of why the algorithm works and be able to expand it in detail for small numbers.

As the students grow older, this experience should lead to increased confidence in understanding these *and similar* algorithms, knowledge of how to construct them in other situations, and the importance of verifying their correctness before relying on them. For example, the process of writing any kind of program for a computer begins with creating algorithms for automating a task and then implementing them on the machine. Without hands-on experience like that described above, students will be ill-equipped to construct correct programs.

Considerations for Grade-Level Accomplishments in Grade Four

The most important mathematical skills and concepts for students in grade four to acquire are described as follows:

- **Multiplication and division facts.** Students who enter the fourth grade without multiplication facts committed to memory are at risk of having difficulty as more complex mathematics is taught. Students' knowledge of basic facts needs to be assessed at the beginning of the school year. Systematic daily practice with multiplication and division facts needs to be provided for students who have not yet learned them.

- **Addition and subtraction.** Mentally adding a two-digit number and a one-digit number is a component skill for working multiplication problems that was targeted in the second grade. Students have to add the carried number to the product of two factors (e.g., 34 × 3). Students should be assessed on the ability to add numbers mentally (e.g., 36 + 7) at the beginning of the school year, and systematic practice should be provided for students not able to work the addition problems mentally.

- **Reading and writing numbers.** Reading and writing numbers in the thousands and millions with one or more zeros in the middle can be particularly troublesome for students. Assessment at the beginning of the fourth grade should test students on reading and writing the more difficult thousand numbers, such as 4,002 and 4,020. When teaching students to read 5- and 6-digit numbers, teachers should be thorough so that students can read, write, and distinguish difficult numbers, such as 300,200 and 320,000.

- **Fractions equal to one.** Understanding fractions equal to one (e.g., 8/8 or 4/4) is important for understanding the procedure for working with equivalent fractions. Students should have an in-depth understanding of how to construct a fraction that equals one to suit the needs of the problem; for example, should a fraction be 32/32 or 17/17? When the class is working on equivalent fraction problems, the teacher should prompt the students on how to find the equivalent fraction or the missing number in the equivalent fraction. The students find the fraction of one that they can use to multiply or divide by to determine the equivalent fraction. (This material is discussed in depth in Appendix A, "Sample Instructional Profile.")

Grade Four

- **Multiplication and division problems.** Multiplication problems in which either factor has a zero are likely to cause difficulties. Teachers should provide extra practice on problems such as 20 × 315 and 24 × 308. Division problems with a zero in the answer may be difficult for students (e.g., 152/3 and 5115/5). Students will need prompting on how to determine whether they have completed the problem of placing enough digits in the answer. (Students who consistently find problems with zeros in the answer difficult to solve may also have difficulties with the concept of place value. Help should be provided to remedy this situation quickly.)

Grade Five | Areas of Emphasis

Chapter 3
Grade-Level
Considerations

By the end of grade five, students increase their facility with the four basic arithmetic operations applied to fractions, decimals, and positive and negative numbers. They know and use common measuring units to determine length and area and know and use formulas to determine the volume of simple geometric figures. Students know the concept of angle measurement and use a protractor and compass to solve problems. They use grids, tables, graphs, and charts to record and analyze data.

Number Sense

1.0   1.1   **1.2**   1.3   **1.4**   **1.5**

2.0   **2.1**   **2.2**   **2.3**   2.4   2.5

Algebra and Functions

1.0   1.1   **1.2**   1.3   **1.4**   **1.5**

Measurement and Geometry

1.0   **1.1**   **1.2**   **1.3**   1.4

2.0   **2.1**   **2.2**   2.3

Statistics, Data Analysis, and Probability

1.0   1.1   1.2   1.3   **1.4**   **1.5**

Mathematical Reasoning

1.0   1.1   1.2

2.0   2.1   2.2   2.3   2.4   2.5   2.6

3.0   3.1   3.2   3.3

Key Standards and Elaboration

By the time students have finished the fourth grade, they should have a basic understanding of whole numbers and some understanding of fractions and decimals. Students at this grade level are expected to have mastered multiplication and division of whole numbers. They should also have had some exposure to negative numbers. These skills will be enhanced in the fifth grade. An important standard focused on enhancing these skills is Number Sense Standard 1.2.

Grade Five

A fraction *c/d* is both "*c* parts of a whole consisting of *d* equal parts" and "the quotient of the number *c* divided by the number *d*."

NUMBER SENSE

1.2 Interpret percents as a part of a hundred; find decimal and percent equivalents for common fractions and explain why they represent the same value; compute a given percent of a whole number.

The fact that a fraction *c/d* is both "*c* parts of a whole consisting of *d* equal parts" and "the quotient of the number *c* divided by the number *d*" was first mentioned in Number Sense Standards 1.5 and 1.6 of grade four. This fact must be *carefully explained* rather than decreed by fiat, as is the practice in most school textbooks. The importance of providing logical explanations for all aspects of the teaching of fractions cannot be overstated because the students' fear of fractions and the mistakes related to them appear to underlie the failure of mathematics education. Once *c/d* is clearly understood to be the division of *c* by *d*, then the conversion of fractions to decimals can be explained logically.

Students will also continue to learn about the relative positions of numbers on the number line, above all, those of negative whole numbers. Negative whole numbers are especially important because, for the first time, they play a major part in core number-sense expectations. Standard 1.5 is important in this regard.

1.5 Identify and represent on a number line decimals, fractions, mixed numbers, and positive and negative integers.

The correct placement of positive fractions on the number line implies that students will need to order and compare fractions. Identifying numbers as points on the real line is an important step in relating students' concept of numbers for arithmetic to geometry. This fusion of arithmetic and geometry, which is ubiquitous in mathematics, adds a new dimension to students' understanding of numbers.

But the most important aspect of students' work with negative numbers is to learn the rules for doing the basic operations of arithmetic with them, as represented in the following standard:

2.1 Add, subtract, multiply, and divide with decimals; add with negative integers; subtract positive integers from negative integers; and verify the reasonableness of the results.

This standard is the beginning of a three-year process of familiarizing students with the full arithmetic of rational numbers. In the fifth grade, students learn how to add negative numbers and how to subtract positive numbers from negative numbers. At this point students should find it profitable to interpret these concepts geometrically. Adding a positive number *b* shifts the point on the number line to the right by *b* units, and adding a negative number −*b* shifts the point on the number line to the left by *b* units, and so forth. Multiplication and division of negative numbers should not be taken up in the fifth grade because division by negative numbers leads to negative fractions, which have not yet been introduced.

The introduction of the general division algorithm is also important, but it can be complicated and consequently difficult for many students to master. In particular, the skills needed to find the largest product of the divisor with an integer between 0 and 9 that is less than the remainder are likely to be demanding for fifth grade students. Students should become comfortable with the algorithm in carefully selected cases in which the numbers needed at each step are clear. Putting such a problem in context may help. For instance, the students might imagine dividing 153 by 25 as packing 153 students into a fleet of buses for a field trip, with each bus carrying a maximum of 25 passengers. Drawing pictures to help with the reasoning, if necessary, can help students to see that it takes six buses with three students left over; those three students get to enjoy being in the seventh bus with room to spare. But it seems both unnecessary and unwise at this stage to extend the concepts beyond what is presented here. The important standard for students to achieve is:

> **2.2** Demonstrate proficiency with division, including division with positive decimals and long division with multidigit divisors.

The most essential number-sense skills that students should learn in the fifth grade are the addition and multiplication of fractions. It should be emphasized that the primary definition of the addition of two fractions a/b and c/d is $ad/bd + bc/bd = (ad + bc)/bd$. This concept should be carefully explained to students instead of being imposed on them without explanation. Otherwise, the common mistakes that students make of believing that $a/b + c/d = (a + c)/(b + d)$ and $a/(b + c) = a/b + a/c$ would be the result. From this definition the usual formula for addition involving the least common multiple of b and d can be easily deduced, but this formula should not be used as *the definition* of adding fractions in general.

Once students have mastered these basic skills with fractions, the number of concrete applications is virtually unlimited. It would be natural to use these concrete problems as drills to promote students' technical fluency with fractions.

Two main skills are involved in reducing fractions: factoring whole numbers in order to put fractions into reduced forms and understanding the basic arithmetic skills involved in this factoring. The two associated standards that should be emphasized are:

> **1.4** Determine the prime factors of all numbers through 50 and write the numbers as the product of their prime factors by using exponents to show multiples of a factor (e.g., $24 = 2 \times 2 \times 2 \times 3 = 2^3 \times 3$).

> **2.3** Solve simple problems, including ones arising in concrete situations involving the addition and subtraction of fractions and mixed numbers (like and unlike denominators of 20 or less), and express answers in the simplest form.

The instructional profile with fractions, which appears later in this chapter, gives many ideas of how to approach this topic. Students may profit from the use of the Sieve of Eratosthenes (see the glossary) in connection with Standard 1.4.

ALGEBRA AND FUNCTIONS

The Algebra and Functions strand for grade five presents one of the key steps in abstraction and one of the defining steps in moving from simply learning arithmetic to learning mathematics: the replacement of numbers by variables.

> **1.2** Use a letter to represent an unknown number; write and evaluate simple algebraic expressions in one variable by substitution.

The importance of this step, which requires *reasoning rather than simple manipulative facility,* mandates particular care in presenting the material. The basic idea that, for example, $3x + 5$ is a shorthand for an infinite number of sums, $3(1) + 5, 3(2.4) + 5, 3(11) + 5$, and so forth, must be thoroughly presented and understood by students; and they must

Grade Five

The primary definition of the addition of two fractions a/b and c/d is $ad/bd + bc/bd = (ad + bc)/bd$.

practice solving simple algebraic expressions. But it is probably a mistake to push too hard here—teachers should not overdrill. Instead, they should check for students' understanding of concepts, perhaps providing students with some simple puzzle problems to give them practice in writing an equation for an unknown from data in a word problem.

Again, in the Algebra and Functions strand, the following two standards are basic:

1.4 Identify and graph ordered pairs in the four quadrants of the coordinate plane.

1.5 Solve problems involving linear functions with integer values; write the equation; and graph the resulting ordered pairs of integers on a grid.

MEASUREMENT AND GEOMETRY

Finally, in Measurement and Geometry these three standards should be emphasized:

1.1 Derive and use the formula for the area of a triangle and of a parallelogram by comparing each with the formula for the area of a rectangle (i.e., two of the same triangles make a parallelogram with twice the area; a parallelogram is compared with a rectangle of the same area by cutting and pasting a right triangle on the parallelogram).

2.1 Measure, identify, and draw angles, perpendicular and parallel lines, rectangles, and triangles by using appropriate tools (e.g., straightedge, ruler, compass, protractor, drawing software).

2.2 Know that the sum of the angles of any triangle is 180° and the sum of the angles of any quadrilateral is 360° and use this information to solve problems.

Students need to *commit to memory* the formulas for the area of a triangle, a parallelogram, and a cube and the formula for the circumference of a circle.

STATISTICS, DATA ANALYSIS, AND PROBABILITY

The ability to graph functions is an essential fundamental skill, and there is no doubt that linear functions are the most important for applications of mathematics. As a result, the importance of these topics can hardly be overestimated. Closely related to these standards are the following two standards from the Statistics, Data Analysis, and Probability strand:

1.4 Identify ordered pairs of data from a graph and interpret the meaning of the data in terms of the situation depicted by the graph.

1.5 Know how to write ordered pairs correctly; for example, (x, y).

These standards indicate the ways in which the skills involved in the Algebra and Functions strand can be reinforced and applied.

Considerations for Grade-Level Accomplishments in Grade Five

At the beginning of grade five, students need to be assessed carefully on their knowledge of the core content taught in the lower grades, particularly in the following areas:

— Knowledge and fluency of basic fact recall, including addition, subtraction, multiplication, and division facts (By this level, students should know all the basic facts and be able to recall them instantly.)

— Mental addition—The ability to mentally add a single-digit number to a two-digit number

— Rounding off numbers in the hundreds and thousands to the nearest ten, hundred, or thousand and rounding off two-place decimals to the nearest tenth
— Place value—The ability to read and write numbers through the millions
— Knowledge of measurement equivalencies, both customary and metric, for time, length, weight, and liquid capacity
— Knowledge of prime numbers and the ability to determine prime factors of numbers up to 50
— Ability to use algorithms to add and subtract whole numbers, multiply a two-digit number and a multidigit number, and divide a multidigit number by a single-digit number
— Knowledge of customary and metric units and equivalencies for time, length, weight, and capacity

All of the topics listed previously need to be taught over an extended period of time. A systematic program must be established to enable students to reach high rates of accuracy and fluency with these skills.

Important mathematical skills and concepts for students in grade five to acquire are as follows:

- **Understanding long division.** Long division requires the application of a number of component skills. Students must be able to round tens and hundreds numbers and work estimation problems, divide a two-digit number into a two- or three-digit number mentally and with paper and pencil, and do the steps in the division algorithm. For grade five it suffices to concentrate on problems in which the estimations give the correct numbers in the quotient. This algorithm needs to be taught efficiently so that excessive amounts of instructional time are not required.

- **Adding and subtracting fractions with unlike denominators.** See the instructional profile (Appendix A) on adding and subtracting fractions with unlike denominators.

- **Working with negative numbers.** The standards call for students to add and subtract negative numbers. Students must be totally fluent with these two operations. Students often become confused with operations with negative numbers because too much is introduced at once, and they do not have the opportunity to master one type before another type is introduced. This material must be presented carefully.

- **Ordering fractions and decimal numbers.** Students can use fraction equivalence skills for comparing fractions and for converting fractions to decimals. Students need to know that 3/4 = 75/100 = 0.75 = 75%.

- **Working with percents.** To compute a given percent of a number, students can convert the percent to a decimal and then multiply. Students must know that 6% translates to 0.06 (percents under ten percent can be troublesome). Students should be assessed on their ability to multiply decimals by whole numbers before work begins on this type of problem.

> Students often become confused with operations with negative numbers because too much is introduced at once.

Grade Six — Areas of Emphasis

By the end of grade six, students have mastered the four arithmetic operations with whole numbers, positive fractions, positive decimals, and positive and negative integers; they accurately compute and solve problems. They apply their knowledge to statistics and probability. Students understand the concepts of mean, median, and mode of data sets and how to calculate the range. They analyze data and sampling processes for possible bias and misleading conclusions; they use addition and multiplication of fractions routinely to calculate the probabilities for compound events. Students conceptually understand and work with ratios and proportions; they compute percentages (e.g., tax, tips, interest). Students know about π and the formulas for the circumference and area of a circle. They use letters for numbers in formulas involving geometric shapes and in ratios to represent an unknown part of an expression. They solve one-step linear equations.

Number Sense

1.0 **1.1** **1.2** **1.3** **1.4**

2.0 2.1 2.2 **2.3** **2.4**

Algebra and Functions

1.0 **1.1** 1.2 1.3 1.4

2.0 2.1 **2.2** 2.3

3.0 3.1 3.2

Measurement and Geometry

1.0 **1.1** 1.2 1.3

2.0 2.1 **2.2** 2.3

Statistics, Data Analysis, and Probability

1.0 1.1 1.2 1.3 1.4

2.0 2.1 **2.2** **2.3** **2.4** **2.5**

3.0 **3.1** 3.2 **3.3** 3.4 **3.5**

Mathematical Reasoning

1.0 1.1 1.2 1.3

2.0 2.1 2.2 2.3 2.4 2.5 2.6 2.7

3.0 3.1 3.2 3.3

Grade Six

Key Standards

NUMBER SENSE

Most of the standards in the Number Sense strand for the sixth grade are very important. These standards can be organized into four groups. The first is the comparison and ordering of positive and negative fractions (i.e., rational numbers), decimals, or mixed numbers and their placement on the number line:

1.1 Compare and order positive and negative fractions, decimals, and mixed numbers and place them on a number line.

Of particular importance is the students' understanding of the positions of the negative numbers and the geometric effect on the numbers of the number line when a number is added or subtracted from them.

The second group is represented by the next three standards, all of which refer to ratios and percents:

1.2 Interpret and use ratios in different contexts (e.g., batting averages, miles per hour) to show the relative sizes of two quantities, using appropriate notations (*a/b*, *a* to *b*, *a:b*).

1.3 Use proportions to solve problems (e.g., determine the value of *N* if 4/7 = *N*/21, find the length of a side of a polygon similar to a known polygon). Use cross-multiplication as a method for solving such problems, understanding it as the multiplication of both sides of an equation by a multiplicative inverse.

1.4 Calculate given percentages of quantities and solve problems involving discounts at sales, interest earned, and tips.

The last standard is especially important because it shows the usefulness of these concepts in real-world situations.

The third group includes the remaining Number Sense standards, all of which relate to fractions:

2.0 Students calculate and solve problems involving addition, subtraction, multiplication, and division.

In view of the discussion in grade five concerning the introduction of negative numbers in grades five through seven, one should concentrate on positive rational numbers (e.g., fractions) in grade six. Helping students to be entirely comfortable with the arithmetic of fractions—to be fluent in the basic operations and *understand* the reasoning behind them—is a critical juncture in their mathematical development.

Within this group the critical skill needed for computing with fractions is to recognize when two fractions are equivalent. The basic standard for this concept follows:

2.4 Determine the least common multiple and the greatest common divisor of whole numbers; use them to solve problems with

Helping students to be entirely comfortable with the arithmetic of fractions is a critical juncture in their mathematical development.

fractions (e.g., to find a common denominator to add two fractions or to find the reduced form for a fraction).

Although the ability to find the greatest common divisor (GCD) of two numbers is needed for deciding whether two fractions are equivalent, no knowledge of the GCD or the least common multiple (LCM) is needed for adding or subtracting two fractions: $a/b + c/d = (ad + bc)/bd$, and $a/b - c/d = (ad - bc)/bd$.

The fourth group stands alone because it consists of only one standard:

2.3 Solve addition, subtraction, multiplication, and division problems, including those arising in concrete situations, that use positive and negative integers and combinations of these operations.

For the first time, students are asked to be completely fluent with the arithmetic of negative integers. Students find this difficult because the reasons for some of the more basic rules seem obscure to them, particularly the formula for multiplying -1 by itself:

$$(-1) \times (-1) = 1$$

A strategy that might be useful in helping students to understand this basic rule is to proceed as follows: First, if a number a satisfies $b + a = 0$; then a is $-b$. That is how $(-b)$ is defined, as the additive inverse of b. Second, $N \times 0 = 0$ for any number N because the area of a rectangle with one side zero is zero. After all, $a \times b$ is the area of a rectangle with height a and length b. So what is the area of a rectangle with height zero? Third:

$$0 = (-1) \times 0 = (-1) \times (1 + (-1)) = (-1) \times (1) + (-1) \times (-1)$$

But this says:

$$(-1) \times (1) = -(-1) \times (-1)$$

Now note that $(-1) \times (1) = (1) \times (-1) = -1$ since (1) multiplied by any number N is just the number N again. Thus the left side of the equation above is -1, so $-1 = -(-1) \times (-1)$ and $(-1) \times (-1)$ must equal 1. The difficulty with an argument like this is that it is very abstract. But it is difficult to produce arguments for this fact that are not abstract.

ALGEBRA AND FUNCTIONS

In the Algebra and Functions strand, the important standards are 1.1 and 2.2. The standard that follows is an expansion of the discussion of linear equations that was begun in the fifth grade:

1.1 Write and solve one-step linear equations in one variable.

The critical importance of these equations for all applied areas of mathematics mandates that students in the sixth grade understand them and be able to solve simple one-variable equations. At a more advanced grade level, students in the sixth grade will be required to solve systems of linear equations. Students also need to justify each step in evaluating linear equations as cited in Standard 1.3. This skill is critical to the algebraic reasoning that is to follow and to the development of carefully applied logic at each step of the process.

Standard 1.1 is closely related to the standards for ratio and percent in the Number Sense strand (Standards 1.2 and 1.4).

 Demonstrate an understanding that *rate* is a measure of one quantity per unit value of another quantity.

Standard 2.2 emphasizes the importance of *understanding* the meaning of the concepts of rate and ratio. Rate and ratio are merely different interpretations in different contexts of dividing one number by another. This standard is also closely

related to the problems of rates, average speed, distance, and time that are introduced in Standard 2.3.

MEASUREMENT AND GEOMETRY

The following core standards are a part of the Measurement and Geometry strand:

- **1.1** Understand the concept of a constant such as π; know the formulas for the circumference and area of a circle.
- **2.2** Use the properties of complementary and supplementary angles and the sum of the angles of a triangle to solve problems involving an unknown angle.

Standard 1.3 is also important, and students should know that the volumes of three-dimensional figures can often be found by dividing and combining them into figures whose volumes are already known.

STATISTICS, DATA ANALYSIS, AND PROBABILITY

The study of statistics is more important in the sixth grade than in the earlier grades. One of the major objectives of studying this topic in the sixth grade is to give students some tools to help them understand the uses and misuses of statistics. The core standards for Statistics, Data Analysis, and Probability that focus on these goals are:

- **2.2** Identify different ways of selecting a sample (e.g., convenience sampling, responses to a survey, random sampling) and which method makes a sample more representative for a population.
- **2.3** Analyze data displays and explain why the way in which the question was asked might have influenced the results obtained and why the way in which the results were displayed might have influenced the conclusions reached.
- **2.4** Identify data that represent sampling errors and explain why the sample (and the display) might be biased.
- **2.5** Identify claims based on statistical data and, in simple cases, evaluate the validity of the claims.

For example, if a study of computer use is focused solely on students from Fresno, the class might try to determine how valid the conclusions might be for the students in the entire state. Again, how valid would the conclusion of a study that interviewed 23 teachers from all over the state be for all the teachers in the state? These questions represent major applications of the type of precise and critical thinking that mathematics is supposed to facilitate in students.

In the sixth grade, students are also expected to become familiar with some of the more sophisticated aspects of probability. They start with the following standard:

- **3.1** Represent all possible outcomes for compound events in an organized way (e.g., tables, grids, tree diagrams) and express the theoretical probability of each outcome.

This strand is challenging but vitally important, not only for its use in statistics and probability but also as an illustration of the power of attacking problems systematically.

The concepts in probability Standards 3.3 and 3.5 may be difficult for students to understand:

- **3.3** Represent probabilities as ratios, proportions, decimals between 0 and 1, and percentages between 0 and 100 and verify that the probabilities computed are reasonable; know that if P is the probability of an event, $1-P$ is the

Grade Six

Students should know that the volumes of three-dimensional figures can often be found by dividing and combining them into figures whose volumes are already known.

probability of an event not occurring.

3.5 Understand the difference between independent and dependent events.

The topics in both standards need to be carefully introduced, and the terms must be defined. Both the concept that probabilities are measures of the likelihood that events might occur (numerical values for probabilities are usually expressed as numbers between 0 and 1) and the distinction between dependent and independent events are important for students to understand. If students can grasp the meaning of the terms, they can understand the basic points of these standards. This knowledge can help students reach accurate conclusions about statistical data.

Considerations for Grade-Level Accomplishments in Grade Six

At the beginning of grade six, students need to be assessed carefully on their knowledge of the core content taught in the early grades, which is described at the beginning of the section for grade five, and on the following content from grade five:

— Increased fluency with the long-division algorithm
— Conversion of percents, decimals, and fractions, including examples that represent a value over 1 (e.g., 2.75 = 2¾ = 275%)
— Use of exponents to show the multiples of a single factor
— Addition, subtraction, multiplication, and division with decimal numbers and negative numbers
— Addition of fractions with unlike denominators and multiplication and division of fractions

All of these topics require teaching over an extended period of time. A systematic program must be established so that students can reach high rates of accuracy and fluency with these skills.

All topics delineated in the grade six standards, and in particular the key strands, should be assessed regularly throughout the sixth grade. Once the skills have been taught and mastery demonstrated through assessment, teachers need to continue to review and maintain the students' skills. Mental mathematics, warm-up activities, and additional questions on tests can be used to accomplish this task.

Important mathematical skills and concepts for students in grade six to acquire are as follows:

- **The least common multiple and the greatest common divisor.** Students can become confused by the concepts of the least common multiple (LCM) and the greatest common divisor (GCD). The least common multiple of two numbers includes examples in which one multiple is in fact the least common multiple (e.g., 2 and 8; the LCM is 8); the least common multiple is the product of the two numbers (e.g., 4 and 5; the LCM is 20); and the least common multiple is a number that fits into neither of the two first categories (6 and 8; the LCM is 24). The teaching sequence should include examples of all three types. Finding the LCM becomes much more difficult with large numbers (e.g., finding the LCM of 36 and 48). One way to determine the answers is with prime factors, $36 = 2 \times 2 \times 3 \times 3$ and $48 = 2 \times 2 \times 2 \times 2 \times 3$. The LCM is $2 \times 2 \times 2 \times 2 \times 3 \times 3$, or 144. The process for finding the LCM can be confused with the process for finding the greatest common divisor (what is the GCD of 12 and 16?) because both deal with multiples of prime factors of numbers. Students should also be told that when a number is very large (e.g., 250 digits), finding its prime factorization is impractical, even with the help of the most powerful computers now available. Thus finding the GCD or LCM of two such large numbers is always possible in theory but can be impossible in practice.

- **Discounts, interest, and tips**. Within this realm are problems that range from simple one-step problems to more complex multistep problems. Programs must be organized so that easier problems are introduced first, followed by a thorough teaching of significantly more difficult problems. An example of a simple discount problem is, *A dress cost 50 dollars. There is a 10 percent discount. How many dollars will the discount be?* This problem is solved by performing the calculation for 10 percent of 50. If the problem asks, *How much will the dress cost with the discount?* the students would have to subtract the discount from the original price. A much more complex problem would be, *The sale price of a dress is 40 dollars. The discount was 20 percent. What was the original cost of the dress?* The problem might be solved through several procedures, all of which would involve the application of many more skills than those called for in the first problem. To work the third problem, the student has to know that the original price equates with 100 percent and the sales price is 80 percent of the original price. One way of solving the problem is for the student to write the equation $0.80 N = 40$, with N representing the original price. Thus $N = 40/0.80 = 50$. This way of solving the problem focuses on the increased emphasis on the use of variables in the Algebra and Functions strand. The computation skills needed to solve for N obviously need to be taught before this type of problem is introduced.

The treatment of interest at this grade is meant to deal with simple interest in one accrual period. It is not intended to extend to compound interest over several accrual periods in which the time is expressed as an exponent, as is the case for the normal computation formula for compound interest.

- **Multiplication and division of fractions**. Students should learn why and how fractions are multiplied and divided. Students must understand why the second fraction in a division problem is inverted, if that process is used. Students need to know when to use multiplication or division in application problems. For example, *There are 24 students in our class. Two-thirds of them passed the test. How many students passed the test?* is solved through multiplying; while the problem, *A piece of cloth that is 12 inches long is going to be cut into strips that are 2/3 of an inch long. How many strips can be made?* is solved through division. Structured systematic teaching must be done to help students determine which procedure to use in solving different problems.

> Students must understand why the second fraction in a division problem is inverted, if that process is used.

Grade Seven — Areas of Emphasis

By the end of grade seven, students are adept at manipulating numbers and equations and understand the general principles at work. Students understand and use factoring of numerators and denominators and properties of exponents. They know the Pythagorean theorem and solve problems in which they compute the length of an unknown side. Students know how to compute the surface area and volume of basic three-dimensional objects and understand how area and volume change with a change in scale. Students make conversions between different units of measurement. They know and use different representations of fractional numbers (fractions, decimals, and percents) and are proficient at changing from one to another. They increase their facility with ratio and proportion, compute percents of increase and decrease, and compute simple and compound interest. They graph linear functions and understand the idea of slope and its relation to ratio.

Number Sense

1.0   1.1   **1.2**   1.3   **1.4**   **1.5**   1.6   **1.7**

2.0   2.1   **2.2**   **2.3**   2.4   **2.5**

Algebra and Functions

1.0   1.1   1.2   **1.3**   1.4   1.5

2.0   2.1   2.2

3.0   3.1   3.2   **3.3**   **3.4**

4.0   **4.1**   **4.2**

Measurement and Geometry

1.0   1.1   1.2   **1.3**

2.0   2.1   2.2   2.3   2.4

3.0   3.1   3.2   **3.3**   **3.4**   3.5   **3.6**

Statistics, Data Analysis, and Probability

1.0   1.1   1.2   **1.3**

Mathematical Reasoning

1.0 1.1 1.2 1.3

2.0 2.1 2.2 2.3 2.4 2.5 2.6 2.7 2.8

3.0 3.1 3.2 3.3

Grade Seven

Negative fractions are formally introduced and studied for the first time.

Key Standards and Elaboration

NUMBER SENSE

The first basic standard for the Number Sense strand is:

 Add, subtract, multiply, and divide rational numbers (integers, fractions, and terminating decimals) and take positive rational numbers to whole-number powers.

At this point the students should understand arithmetic involving rational numbers. Negative fractions are formally introduced and studied for the first time. They should know the difference between rational and irrational numbers (Standard 1.4) and be aware that numbers such as the square root of two are not rational. Here, teachers should take care not to misinform the students. For example, some textbooks assert that the square root of 2 is not a rational number and then "prove" that assertion by producing a calculator-generated representation of $\sqrt{2}$ to perhaps 15 decimal places and state that the decimal is not repeating. That is unacceptable. It is better to use the facts in the standard (Standard 1.5) to construct an explicit nonrepeating decimal:

1.5 Know that every rational number is either a terminating or a repeating decimal and be able to convert terminating decimals into reduced fractions.

One can construct a nonrepeating decimal, for example, by putting zeros in all the places past the decimal point except for those in (1) the first, second, fourth, and eighth places and, generally, each power of two:

0.1101000100000001000000000000000010000 . . .

or perhaps (2) the first, third, sixth, tenth, and, generally, the $\dfrac{n(n+1)}{2}$ position

0.101001000100001000001000000100 . . . .

In this way students will see how to construct vast quantities of irrational numbers. At this point it might be possible to challenge the advanced students by showing them that a specific number (such as $\sqrt{2}$) is, in fact, irrational. They then can learn that while there are vast quantities of both rational and irrational numbers, it is often very difficult to show that specific numbers are in one set or the other. But this sophisticated material should not be emphasized for the class as a whole. In particular, at this stage it is probably not wise to attempt any kind of a proof of the facts in Standard 1.5. The students can be told that this basic awareness of irrationality is sufficiently important to be discussed at this point even though its justification will have to be deferred until they take a more advanced course.

By now the students should have enough skill with factoring integers so that they can use factoring to find the smallest common denominator of two whole numbers (Standard 2.2). Teachers should emphasize, once again, that the correct definition of the sum of two fractions is $(a/b) + (c/d) = (ad + bc)/bd$ and that the usual algorithm using factoring to find the smallest common denominator is but a refinement of the primary definition. (See the discussion in the Number Sense standards for the fifth grade.) For this topic students should become more familiar with the basic exponent rules (Standard 2.3), which will have direct applications in the main seventh grade application of compound interest.

The last topic in the first standard of the Number Sense strand (Standard 1.7) is also one of the high points of the entire strand:

> **1.7** Solve problems that involve discounts, markups, commissions, and profit and compute simple and compound interest.

This is a major topic, which should come toward the end of the year and should be a major highlight of the kindergarten through grade seven mathematical experience. It provides one of the most important applications of mathematics in students' everyday life, a skill that can mean the difference between students managing their money and other resources well or not at all.

Standard 2.5, the last standard in the Number Sense strand, on absolute value should receive some emphasis. This topic is usually slighted in middle schools and high schools; however, students should acquire some facility with this concept as early as possible. The students need to understand that the correct way to express the statement "two numbers x and y are close to each other" is "$|x-y|$ is small." The concept of two numbers being "close" was introduced in grade four in connection with rounding off (see "Elaboration" in grade four).

ALGEBRA AND FUNCTIONS

Familiarity with the distributive law, the associative law, and the commutative rule for addition and multiplication of whole numbers has been mentioned at several points previously in the Algebra and Functions standards in grades five and six. For these standards in grade seven, the concepts are taken a step further with the following:

> **1.3** Simplify numerical expressions by applying the properties of rational numbers (e.g., identity, inverse, distributive, associative, commutative) and justify the process used.

This is a critical step in learning how to abstract and shows the power of abstract thinking in helping to make sense of complex situations and derive their basic properties.

One of the most basic topics in applications of mathematics is systems of linear equations. A clear understanding of even something as simple as systems of two linear equations in two unknowns is crucial to understanding more advanced topics, such as calculus and analysis. The first major steps are taken toward this goal when the study of a single linear equation is initiated in these four standards:

> **3.3** Graph linear functions, noting that the vertical change (change in y-value) per unit of horizontal change (change in x-value) is always the same and know that the ratio ("rise over run") is called the slope of a graph.

> **3.4** Plot the values of quantities whose ratios are always the same (e.g., cost to the number of an item, feet to inches, circumference to diameter of a circle). Fit a line to the plot and

understand that the slope of the line equals the quantities.

4.1 Solve two-step linear equations and inequalities in one variable over the rational numbers, interpret the solution or solutions in the context from which they arose, and verify the reasonableness of the results.

4.2 Solve multistep problems involving rate, average speed, distance, and time or a direct variation.

Again, the connection of the second standard with the Measurement and Geometry Standard 1.3 should be noted. These topics provide excellent problems to test the students' understanding of the techniques for solving linear equations.

Students at this stage of algebraic development should be able to understand a clarification of the somewhat subtle concepts of *ratio* and *direct variation* (sometimes called *direct proportion*). The "ratio between two quantities" is nothing more or less than a particular interpretation of "one quantity divided by another in the sense of numbers." Of course, thus far students know only how to divide rational numbers. The teacher should tell the students that the division between irrational numbers will also be explained to them in more advanced courses; therefore, this definition of *ratio* will still apply. *Direct variation* can be explained in terms of linear functions: "A varies directly with B" means that "for a fixed constant c, $A = cB$." Teachers and textbooks commonly try to "explain" the meanings of both terms in abstruse language, resulting in confusion among students and even teachers. No explanation is necessary: *ratio* and *direct variation* are mathematical terms, and they should be clearly defined once the students have been taught the necessary facts and techniques.

MEASUREMENT AND GEOMETRY

The first major emphasis in the Measurement and Geometry strand is for the students to develop an increased sense of spatial relations. This topic is reflected in these two standards:

3.4 Demonstrate an understanding of conditions that indicate two geometrical figures are congruent and what congruence means about the relationships between the sides and angles of the two figures.

3.6 Identify elements of three-dimensional geometric objects (e.g., diagonals of rectangular solids) and describe how two or more objects are related in space (e.g., skew lines, the possible ways three planes might intersect).

A critical part of understanding this material is that the students know the general definition of *congruence*—two figures are congruent if they can be identified by a succession of reflections, rotations, and translations—and understand that properties of two congruent figures, such as angles, edge lengths, areas, and volumes, are equal.

The next basic step is contained in the following standard:

3.3 Know and understand the Pythagorean theorem and its converse and use it to find the length of the missing side of a right triangle and the lengths of other line segments and, in some situations, empirically verify the Pythagorean theorem by direct measurement.

The Pythagorean theorem is probably the first true theorem that the students will have seen. It should be emphasized that students are not expected to prove this result. But the better students should

Chapter 3
Grade-Level Considerations

Grade Seven

The Pythagorean theorem is probably the first true theorem that the students will have seen.

Grade Seven

Seventh grade students should memorize the formulas for the volumes of cylinders and prisms.

be able to understand the proof given by cutting, in two different ways, a square with the edges of length $a + b$ (where a and b are the lengths of the legs of the right triangle). However, everyone is expected to understand what the theorem and its converse mean and how to use both. The applications can include understanding the formula that the square root of $x^2 + y^2$ is the length of the line segment from the origin to the point (x, y) in the plane and that the shortest distance from a point to a line not containing the point is the length of the line segment from the point perpendicular to the line.

Although the following topics are not as basic as the preceding ones, they should also be covered carefully. Seventh grade students should memorize the formulas for the volumes of cylinders and prisms (Standard 2.1). Students at this point should understand the discussion that began in the sixth grade concerning the volume of "generalized cylinders." More precisely, they should think of a right circular cylinder as the solid traced by a circular disc as this disc moves up a line segment L perpendicular to the disc itself. The disc is replaced with a planar region of any shape, and the line segment L is no longer required to be perpendicular to the planar region. Then, as the planar region moves up along L, always parallel to itself, it traces out a solid called a generalized cylinder. The formula for the volume of such a solid is still (height of the generalized cylinder) × (area of the planar region). *Height* now refers to the vertical distance between the top and bottom of the generalized cylinder.

The final topic to be emphasized in seventh grade Measurement and Geometry is as follows:

> **1.3** Use measures expressed as rates (e.g., speed, density) and measures expressed as products (e.g., person-days) to solve problems; check the units of the solutions; and use dimensional analysis to check the reasonableness of the answer.

This standard interacts well with the demands of the algebra standards, particularly in solving linear equations. Typically, the main difficulty in understanding problems of this kind is keeping the concepts straight; therefore, care should be taken to emphasize the meanings of the terms involved in the various problems.

STATISTICS, DATA ANALYSIS, AND PROBABILITY

The most important of the three seventh grade standards in Statistics, Data Analysis, and Probability is this:

> **1.3** Understand the meaning of, and be able to compute, the minimum, the lower quartile, the median, the upper quartile, and the maximum of a data set.

These are useful measures that students need to know well. Care should be taken to ensure that all students know the definitions, and many examples should be given to illustrate them.

Preface to Grades Eight Through Twelve

The standards for grades eight through twelve are organized differently from those for kindergarten through grade seven. (A complete description of this organization is provided on page 72, "Introduction to Grades Eight Through Twelve.") In grades eight through twelve, the mathematics studied is organized according to disciplines such as algebra and geometry. *Local educational agencies may choose to teach high school mathematics in a traditional sequence of courses (Algebra I, geometry, Algebra II, and so forth) or in an integrated fashion in which some content from each discipline is taught each year.*

However mathematics courses are organized, the core content of these subjects must be covered by the end of the sequences of courses, and all academic standards for achievement must be the same. The core content and the emphasis areas are delineated in the discussions of the individual disciplines presented in this section.

What follows in this preface is a discussion of key standards and discipline-level emphases for Algebra I, geometry, Algebra II, and probability and statistics. These same disciplines will be tested under the statewide Standardized Testing and Reporting (STAR) program, which will offer both traditional discipline-based versions and integrated versions of its test. The following section describes standards for the academic content by discipline, along with the areas of emphasis in each discipline; it is not an endorsement of a particular choice of structure for courses or a particular method of teaching the mathematical content. The additional advanced subjects of mathematics covered in the standards (linear algebra, advanced placement probability and statistics, and calculus) are not discussed in this section because many of these advanced subjects are not taught in every middle school or high school. Schools and districts may combine the subject matter of these various disciplines. Many combinations of these subjects are possible, and this framework does not prescribe a single instructional approach.

By the eighth grade, students' mathematical sensitivity should be sharpened. Students should start perceiving logical subtleties and appreciating the need for sound mathematical arguments before making conclusions. As students progress in the study of mathematics, they learn to understand the meaning of logical implication; test general assertions; realize that one counterexample is enough to show that a general assertion is false; conceptually understand that the truth of a general assertion in a few cases does not allow the

Chapter 3
Grade-Level Considerations

However mathematics courses are organized, all academic standards for achievement must be the same.

conclusion that it is true in all cases; distinguish between something being proven and a mere plausibility argument; and identify logical errors in chains of reasoning.

From kindergarten through grade seven, these standards have impressed on the students the importance of logical reasoning in mathematics. Starting with grade eight, students should be ready for the basic message that logical reasoning is the underpinning of all mathematics. In other words, every assertion can be justified by logical deductions from previously known facts. Students should begin to learn to *prove* every statement they make. Every textbook or mathematics lesson should try to convey this message and to convey it well.

[For information on problems from the Third International Study of Mathematics and Science (TIMSS), readers are referred to a resource kit, *Attaining Excellence: A TIMSS Resource Kit,* and to a Web site < http://www.csteep.bc.edu/TIMSS1/pubs_main.html>.]

Mathematical Proofs

A misapprehension in mathematics education is that proofs occur only in Euclidean geometry and that elsewhere one merely learns to *solve problems* and *do computations*. Problem solving and symbolic computations are nothing more than different manifestations of mathematical proofs. To illustrate this point,

the following discussion shows how the usual computations leading to the solution of a simple linear equation are nothing but the steps of a well-disguised proof of a theorem.

Consider the problem of solving this equation:

$$x - \tfrac{1}{4}(3x - 1) = 2x - 5$$

Multiply both sides by 4 to get:

$$4x - (3x - 1) = 8x - 20$$

Then simplify the left side to get:

$$x + 1 = 8x - 20$$

Transposing x from left to right yields:

$$1 = 7x - 20$$

One more transposition gives the result $x = 3$.

So far this seems to be an entirely mechanical procedure. No proof is involved.

Closer examination reveals that what is really being stated is a mathematical theorem:

A number x satisfies
$x - \tfrac{1}{4}(3x - 1) = 2x - 5$
when and only when $x = 3$.

That $x = 3$ satisfies the equation is easy to see. The less trivial part of the preceding theorem is the assertion that if a number x satisfies $x - \tfrac{1}{4}(3x - 1) = 2x - 5$, then x is necessarily equal to 3. A *proof* of this fact is presented next in a two-column format:

1. $x - \frac{1}{4}(3x - 1) = 2x - 5$	1. Hypothesis
2. $4(x - \frac{1}{4}(3x - 1)) = 4(2x - 5)$	2. $a = b$ implies $ca = cb$ for all numbers a, b, c.
3. $4x - 4(\frac{1}{4}(3x - 1)) = 4(2x) - 20$	3. Distributive law
4. $4x - (4 \cdot \frac{1}{4})(3x - 1) = (4 \cdot 2)x - 20$	4. Associative law for multiplication
5. $4x - (3x - 1) = 8x - 20$	5. $1 \cdot a = a$ for all numbers a
6. $4x + (-3x + 1) = 8x - 20$	6. $-(a - b) = (-a + b)$ for all numbers a, b.
7. $(4x + (-3x)) + 1 = 8x - 20$	7. Associative law for addition
8. $x + 1 = 8x - 20$	8. $4x + (-3x) = (4 + (-3))x$, by the distributive law
9. $-x + (x + 10) = (-x + 8x) - 20$	9. Equals added to equals are equal.
10. $(-x + x) + 1 = (-x + 8x) - 20$	10. Associative law for addition: $0 + 1 = 1$.
11. $1 = 7x - 20$	11. $-x + 8x = (-1 + 8)x$, by the distributive law
12. $1 + 20 = (7x - 20) + 20$	12. Equals added to equals are equal.
13. $21 = 7x + [(-20) + 20]$	13. Associative law for addition
14. $21 = 7x$	14. $-a + a = 0$ for all a; $b + 0 = b$ for all b.
15. $3 = x$	15. Multiply (14) by $\frac{1}{7}$ and apply the associative law to $\frac{1}{7}(7x)$.
16. $x = 3$	16. $a = b$ implies that $b = a$ Q.E.D.

In practice, it would be impractical to demand such detail each time a linear equation is solved. Nevertheless, without the realization that such a mathematical proof is lurking behind the well-known formalism of solving linear equations, an author of an algebra textbook or a teacher in a classroom would most likely emphasize the wrong points in the presentation of beginning algebra.

The preceding proof clearly exposes the need for *generality* in the presentation of the associative laws and distributive law. In these standards these laws are taught starting with grade two, but it is probably difficult to convince students that such seemingly obvious statements deserve discussion. For example, if one has to believe that $3(5 + 11) = 3 \cdot 5 + 3 \cdot 11$, all one has to do is to expand both sides:

Chapter 3
Grade-Level Considerations

Grades Eight Through Twelve

Without the realization that a mathematical proof is lurking behind the well-known formalism of solving linear equations, a teacher would most likely emphasize the wrong points in the presentation of beginning algebra.

Chapter 3
Grade-Level Considerations

Grades Eight Through Twelve

clearly, $3 \cdot 16 = 15 + 33$ because both sides are equal to 48. However, one look at the deduction of step 3 from step 2 in the preceding mathematical demonstration would make it clear that the hands-on approach to the distributive law is useless in this situation. Begin with the right-hand side of the equation:

$$4(2x - 5) = 4(2x) - 4 \cdot 5$$

Here x is an *arbitrary* number, so we are not saying that

$$4(2 \cdot 17 - 5) = 4(2 \cdot 17) - 4 \cdot 5$$

or that

$$4(2 \cdot 172 - 5) = 4(2 \cdot 172) - 4 \cdot 5$$

Were that the case, the equality could again have been verified by expanding both expressions. Rather, the assertion is that, *although we do not know what number x is*, nevertheless it is true that $4(2x - 5) = 4(2x) - 4 \cdot 5$. There is no alternative except to justify this general statement by using a general rule: the distributive law. The same comment applies to the other applications of the associative laws and the distributive law in the preceding proof.

It must be recognized that some proofs may not be accessible until the later grades, such as the reason for the formula of the circumference of a circle, $C = 2\pi r$. Nevertheless, *every* technique taught in mathematics is nothing but proofs in disguise. The validity of this statement can be revealed by considering a special case, such as this word problem for grade eight:

Jan had a bag of marbles. She gave one-half to James and then one-third of the marbles still in the bag to Pat. She then had 6 marbles left. How many marbles were in the bag to start with? (TIMSS, gr. 8, N-16)

The solution to the problem follows:

Suppose Jan had n marbles to start with. If she gave one-half to James, then she had $n/2$ marbles left. According to the problem, she then gave one-third of what was left to Pat (i.e., she gave $(⅓) \cdot (½)n$ to Pat). Thus she gave $(⅙)n$ marbles to Pat, and what she had left was $(½)n - (⅙)n = (⅓)n$. But the problem states that Jan had "6 marbles left." So $(⅓)n = 6$, and $n = 18$. Therefore, Jan had 18 marbles to begin with.

The next step is to analyze in what sense the preceding solution masks a proof. First, the usual solution as presented previously can be broken into two distinct steps:

1. *Setting up the equation:* If n is the number of marbles Jan had to begin with, then the given data imply:
   $$(n - (½)n) - (⅓)(n - (½)n) = 6$$

2. *Solving the equation:* This step requires the proof of the following theorem:
   n satisfies the equation $(½)n - (⅓)(½)n = 6$ when and only when $n = 18$.

Step 1 and step 2 exemplify the two components of mathematics in grades eight through twelve: teaching the skills needed to transcribe sometimes untidy raw data into mathematical terms and teaching the skills needed to draw precise logical conclusions from clearly stated hypotheses. Neither can be slighted.

It should be pointed out, however, that the built-in uncertainty and indeterminacy of step 1—which can lead to the setting up of several distinct equations and hence several distinct solutions—has led to the view of mathematics as an imprecise discipline in which a problem may have more than one correct answer. *This lack of understanding of the sharp distinction between step 1 and step 2 has had the deleterious effect of downgrading the importance of obtaining a single correct* answer and jettisoning the inherent precision *of mathematics*. As a result the rigor and precision needed for step 2 have been vigorously questioned. Such a misconception of mathematics would never have materialized had the process of transcription been better understood. This level of rigor and precision is embedded in the standards and is essential so that all students can develop mathematically to the level required in the *Mathematics Content Standards*.

Misconceptions in Mathematics Problems

The following is an extreme example of the kind of misconception discussed earlier:

> *The 20 percent of California families with the lowest annual earnings pay an average of 14.1 percent in state and local taxes, and the middle 20 percent pay only 8.8 percent. What does that difference mean? Do you think it is fair? What additional questions do you have?*

The preceding type of problem is typically presented as a school mathematics problem. However, a proper understanding of the difference in the two figures of 14.1 percent and 8.8 percent would require a strong background in politics, economics, and sociology. Most students do not have this kind of knowledge. Moreover, the idea of "fairness" is a difficult one even for professional political scientists and sociologists. Formulating a *mathematical* transcription of this elusive concept in this context is therefore beyond the grasp of the best professionals, much less that of school students. Since it is impossible to transcribe the problem into mathematics, step 1 (setting up the equation) cannot be carried out, so there can be no step 2 (solving the equation). This is therefore not a mathematical problem. Hence, the fact that it has no single correct answer can in no way lend credence to the assertion that mathematics is uncertain or imprecise.

The preceding discussion explains that mathematical proofs are the underpinning of all of mathematics. Beginning with grade eight, students must deepen their understanding of the essential foundations for reasoning provided by mathematical proofs. It would be counterproductive to force every student to write a two-column proof at every turn, and it would be equally foolish to require all mathematics instructional materials to be as pedantic about giving such details as the two-column proof shown earlier in this preface. Nevertheless, the message that proofs underlie everything being taught should be clear in the instructional material and mathematical lessons taught in grades eight through twelve. In particular, all instructional materials—not just those for geometry, but especially those for algebra and trigonometry—should carefully present proofs of mathematical assertions when the situation calls for them. For example, an algebra textbook which asserts that a polynomial $p(x)$ satisfying $p(a) = 0$ for some number a must contain $x - a$ as a factor, but which does not offer a detailed proof beyond a few concrete examples for corroboration, is not presenting material compatible with the standards.

Grades Eight Through Twelve

Mathematical proofs are the underpinning of all of mathematics.

Algebra I

In algebra, students learn to reason symbolically, and the complexity and types of equations and problems that they are able to solve increase dramatically as a consequence. The key content for the first course, Algebra I, involves understanding, writing, solving, and graphing linear and quadratic equations, including systems of two linear equations in two unknowns. Quadratic equations may be solved by factoring, completing the square, using graphs, or applying the quadratic formula. Students should also become comfortable with operations on monomial and polynomial expressions. They learn to solve problems employing all of these techniques, and they extend their mathematical reasoning in many important ways, including justifying steps in an algebraic procedure and checking algebraic arguments for validity.

Transition from Arithmetic to Algebra

Perhaps the fundamental difficulty for many students making the transition from arithmetic to algebra is their failure to recognize that the symbol x stands for a number. For example, the equation $3(2x - 5) + 4(x - 2) = 12$ simply means that a certain number x has the property that when the arithmetic operations $3(2x - 5) + 4(x - 2)$ are performed on it as indicated, the result is 12. The problem is to find that number (solution). Teachers can emphasize this point by having students perform a series of arithmetic computations (using pen and paper) starting with $x = 1$, $x = 2$, $x = 3$, $x = 4$, and so forth, thereby getting -13, -3, 7, 17, and so forth. These computations show that none of 1, 2, 3, 4 can be that solution. Going from $x = 3$ to $x = 4$, the value of the expression changes from 7 to 17; therefore, it is natural to guess that the solution would be between 3 and 4. More experimentation eventually gives 3.5 as the solution.

Working backwards, since $3(2(3.5) - 5) + 4((3.5) - 2) = 12$, one can apply the distributive law and commutative and associative laws to unwind the expression, intentionally not multiplying out $2(3.5)$, $4(3.5)$, and so forth, to get:

$$3.5 = \frac{12 + 3(5) + 4(2)}{6 + 4}$$

But this is exactly the principle of solving the equation $3(2x - 5) + 4(x - 2) = 12$ for the number x:

$$x = \frac{(12 + 3(5) + 4(2))}{(6 + 4)}$$

One can bring closure to such a lesson by stressing the similarity between the handling of the algebraic equation and the earlier simple arithmetic operations.

Basic Skills for Algebra I

The first basic skills that must be learned in Algebra I are those that relate to understanding linear equations and solving systems of linear equations. In Algebra I the students are expected to solve only two linear equations in two unknowns, but this is a basic skill. The following six standards explain what is required:

4.0 Students simplify expressions before solving linear equations and inequalities in one variable, such as $3(2x - 5) + 4(x - 2) = 12$.

5.0 Students solve multistep problems, including word problems, involving linear equations and linear inequalities in one variable and provide justification for each step.

6.0 Students graph a linear equation and compute the *x*- and *y*-intercepts (e.g., graph $2x + 6y = 4$). They are also able to sketch the region defined by linear inequality (e.g., they sketch the region defined by $2x + 6y < 4$).

7.0 Students verify that a point lies on a line, given an equation of the line. Students are able to derive linear equations by using the point-slope formula.

9.0 Students solve a system of two linear equations in two variables algebraically and are able to interpret the answer graphically. Students are able to solve a system of two linear inequalities in two variables and to sketch the solution sets.

15.0 Students apply algebraic techniques to solve rate problems, work problems, and percent mixture problems.

Each of these standards can be a source of difficulty for students, but they all reflect basic skills that must be understood so that students can advance to the next level in their understanding of mathematics. Moreover, modern applications of mathematics rely on solving systems of linear equations more than on any other single technique that students will learn in kindergarten through grade twelve mathematics. Consequently, it is essential that they learn these skills well.

Point-Slope Formula

Perhaps the most perplexing difficulty that students have is with Standard 7.0. It often seems very hard for them to understand this point. But it is one of the most critical skills in this section. In particular, the following idea must be clearly understood before the students can progress further: A point lies on a line given by, for example, the equation $y = 7x + 3$ if and only if the coordinates of that point (a, b) satisfy the equation when x is replaced with a and y with b. One way of explaining this idea is to emphasize that the graph of the equation $y = 7x + 3$ is precisely the set of points (a, b) for which replacing x by a and y by b gives a true statement. (For example, $(3, 2)$ is not on the graph because replacing x with 3 and y with 2 gives the statement $2 = 23$, which is not true.) Thus, the graph consists of all points of the form $(a, 7a + 3)$. It also follows from these considerations that the root r of the linear polynomial $7x + 3$ is the *x*-intercept of the graph of $y = 7x + 3$ because $(r, 0)$ is on the graph.

An additional comment about Standard 7.0 is that, although it singles out the point-slope formula, it is understood that students also have to know how to write the equation of a line when two of its points are given. However, the fact that the slope of a line is the same regardless of which pair of points on the line are used

Modern applications of mathematics rely on solving systems of linear equations more than on any other single technique.

for its definition depends on the considerations of similar triangles. (This fact is first mentioned in Algebra and Functions Standard 3.3 for grade 7.) This small gap in the logical development should be made clear to students, with the added assurance that they will learn the concept in geometry. The same comment applies also to the fact that two lines are perpendicular if and only if the product of their slopes is −1 (Standard 8.0).

Quadratic Equations

The next basic topic is the development of an understanding of the structure of quadratic equations. Here, one repeats the considerations involved in linear equations, such as graphing and understanding what it means for a point (x, y) to be on the graph. In particular, the graphical interpretation of finding the zeros of a quadratic equation by identifying the x-intercepts with the graph is very important and, as was the case with linear equations, is also a source of serious difficulty. Equally important is the recognition that if a, b are the roots of a quadratic polynomial, then up to a multiplicative constant, it is equal to $(x - a)(x - b)$.

When the discriminant of a quadratic polynomial is negative, the quadratic formula yields no information at this point because students have not yet been introduced to complex numbers. This deficiency will be remedied in Algebra II. The following standards show which skills students in a first-year algebra course need for solving quadratic equations:

- **14.0** Students solve a quadratic equation by factoring or completing the square.
- **19.0** Students know the quadratic formula and are familiar with its proof by completing the square.
- **20.0** Students use the quadratic formula to find the roots of a second-degree polynomial and to solve quadratic equations.
- **21.0** Students graph quadratic functions and know that their roots are the x-intercepts.
- **23.0** Students apply quadratic equations to physical problems, such as the motion of an object under the force of gravity.

Additional Comments

Students should be carefully guided through the solving of word problems by using symbolic notations. Many students may be so overwhelmed by the symbolic notation that they start to manipulate symbols carelessly, and word problems become incomprehensible. Teachers and publishers need to be sensitive to this difficulty. In addition to Standard 15.0, cited previously, the other relevant standards for solving word problems using symbolic notations are:

- **10.0** Students add, subtract, multiply and divide monomials and polynomials. Students solve multistep problems, including word problems, by using these techniques.
- **13.0** Students add, subtract, multiply, and divide rational expressions and functions. Students solve both computationally and conceptually challenging problems by using these techniques.

Among the word problems of this level, those involving *direct* and *inverse* proportions occupy a prominent place. These concepts, which are often mired in the language of "proportional thinking," need clarification. A quantity P is said to be *proportional to* another quantity Q if the quotient P/Q is a fixed constant k. This k is then called the *constant of proportional-*

ity. Students should be made aware that this is a mathematical definition, and there is no need to look for linguistic subtleties concerning the phrase "to be proportional to." Similarly, P is said to be *inversely proportional* to Q if the product PQ is equal to a fixed constant h.

In Standard 13.0 the emphasis should be on *formal* rational expressions instead of on rational *functions*. Many of these formal techniques will become increasingly important in Algebra II and trigonometry. The rules of exponents, for example, are fundamental to an understanding of the exponential and logarithmic functions. Many students fail to cope with the latter topics because their understanding of the rules of (fractional) exponents is weak. The skills in the following standards need to be emphasized in a first-year algebra course:

2.0 Students understand and use such operations as taking the opposite, finding the reciprocal, taking a root, and raising to a fractional power. They understand and use the rules of exponents.

12.0 Students simplify fractions with polynomials in the numerator and denominator by factoring both and reducing them to the lowest terms.

The gist of Standards 16.0 through 18.0 is to introduce students to a precise concept of functions in the language of ordered pairs. Introducing this concept needs to be done carefully because students at this stage of their mathematical development may not be ready for this level of abstraction. However, during a first-year algebra course is the stage at which students should see and use the functional notation $f(x)$ for the first time.

In Standard 24.0 students begin to learn simple logical arguments in algebra. They can be taught the proof that square roots of prime numbers are never rational, thereby solidifying to a certain extent their understanding of rational and irrational numbers (grade seven, Number Sense Standard 1.4). In Standard 3.0 students are taught to solve equations and inequalities involving absolute values, but it is not necessary to introduce the interval notation $[a, b]$, (a, b), $[a, b)$, and so forth at this point. However, they should be introduced to the set notation $\{a, b, c, \ldots\}$ and $\{x : x \text{ satisfies property } P\}$ and to the empty set ϕ in, for example, Standard 17.0. Finally, students should become familiar with the terminology "solution set" of Standard 9.0—meaning the set of all solutions.

Chapter 3
Grade-Level
Considerations

Grades Eight
Through
Twelve

Algebra I

During a first-year algebra course is the stage at which students should see and use the functional notation $f(x)$ for the first time.

Geometry

> Students should be encouraged to draw many pictures to develop a geometric sense and to amass a wealth of geometric data in the process.

The main purpose of the geometry curriculum is to develop geometric skills and concepts and the ability to construct formal logical arguments and proofs in a geometric setting. Although the curriculum is weighted heavily in favor of plane (synthetic) Euclidean geometry, there is room for placing special emphasis on coordinated geometry and its transformations.

The first standards introduce students to the basic nature of logical reasoning in mathematics:

1.0 Students demonstrate understanding by identifying and giving examples of undefined terms, axioms, theorems, and inductive and deductive reasoning.

3.0 Students construct and judge the validity of a logical argument and give counterexamples to disprove a statement.

Starting with undefined terms and axioms, students learn to establish the validity of other assertions through logical deductions; that is, they learn to prove theorems. This is their first encounter with an axiomatic system, and experience shows that they do not easily adjust to the demand of total precision needed for the task. In general, it is important to impress on students from the beginning that the main point of a proof is the mathematical correctness of the argument, not the literary polish of the writing or the adherence to a particular proof format.

Inductive Reasoning

Standard 1.0 also calls for an understanding of inductive reasoning. Students are expected not only to recognize inductive reasoning in a formal sense but also to demonstrate how to put it to use. To this end students should be encouraged to draw many pictures to develop a geometric sense and to amass a wealth of geometric data in the process. Many students—including high-achieving ones—complete a course in geometry with so little geometric intuition that, given three noncollinear points, they cannot even begin to visualize what the circumcircle of these points must be like. One way to develop this geometric sense is to have the students become familiar with the basic straightedge-compass constructions, as illustrated in the following standard:

16.0 Students perform basic constructions with a straightedge and compass, such as angle bisectors, perpendicular bisectors, and the line parallel to a given line through a point off the line.

It would be desirable to introduce students to these constructions early in the course and leave the proofs of their validity to the appropriate place of the logical development later.

Geometric Proofs

The subject then turns to geometric proofs in earnest. The foundational results of plane geometry are embodied in the following standards:

- **2.0** Students write geometric proofs, including proofs by contradiction.
- **4.0** Students prove basic theorems involving congruence and similarity.
- **7.0** Students prove and use theorems involving the properties of parallel lines cut by a transversal, the properties of quadrilaterals, and the properties of circles.
- **12.0** Students find and use measures of sides and of interior and exterior angles of triangles and polygons to classify figures and solve problems.
- **21.0** Students prove and solve problems regarding relationships among chords, secants, tangents, inscribed angles, and inscribed and circumscribed polygons of circles.

It has become customary in high school geometry textbooks to start with axioms that incorporate real numbers. Although doing geometric proofs with real numbers runs counter to the spirit of Euclid, this approach is a good mathematical compromise in the context of school mathematics. However, the parallel postulate occupies a special place in geometry and should be clearly stated in the traditional form: Through a point not on a given line L, there is at most one line parallel to L. Because this postulate played a fundamental role in the development of mathematics up to the nineteenth century, the significance of the postulate should be discussed. And because there always exists at least one parallel line through a point to a given line, the import of this postulate lies in the uniqueness of the parallel line.

A discussion of this postulate provides a natural context to show students the key concept of uniqueness in mathematics—a concept that experience indicates students usually find difficult.

It is also recommended that the topics of circles and similarity be taught as early as possible. Once those topics have been presented, the course enters a new phase not only because of the interesting theorems that can now be proved but also because the concept of similarity expands the applications of algebra to geometry. These applications might include determining one side of a regular decagon on the unit circle through the use of the quadratic formula as well as the applications of geometry to practical problems.

It is often not realized that theorems for circles can be introduced very early in a geometry course. For instance, the remarkable theorem that inscribed angles on a circle which intercept equal arcs must be equal can in fact be presented within three weeks after the introduction of axioms. All it takes is to prove the following two theorems:

1. Base angles of isosceles triangles are equal.
2. The exterior angle of a triangle equals the sum of opposite interior angles.

At this point it is necessary to deal with one of the controversies in mathematics education concerning the format of proofs. It has been argued that the traditional two-column format is stultifying for students and that the format for proofs in the mathematics literature is always paragraph proofs. While the latter observation is true, teachers should be aware that a large part of the reason for using paragraph proofs is the expense of typesetting more elaborate formats, not that paragraph proofs are intrinsically better or clearer. In fact, neither of these claims of superiority for paragraph proofs is actually valid. Furthermore, it appears that for beginners to learn the precision of argument needed, the two-column format is best. After the students have shown a mastery of the basic logical skills, it would be appropriate to relax the requirements on form. *But the teacher should never relax the requirement that all arguments presented by the students be precise and correct.*

Pythagorean Theorem

One of the high points of elementary mathematics, in fact of all of mathematics, is the Pythagorean theorem:

> **14.0** Students prove the Pythagorean theorem.

This theorem can be proved initially by using similar triangles formed by the altitude on the hypotenuse of a right triangle. Once the concept of area is introduced (Standard 8.0), students can prove the Pythagorean theorem in at least two more ways by using the familiar picture of four congruent right triangles with legs a and b nestled inside a square of side $a + b$.

> **8.0** Students know, derive, and solve problems involving the perimeter, circumference, area, volume, lateral area, and surface area of common geometric figures.
>
> **10.0** Students compute areas of polygons, including rectangles, scalene triangles, equilateral triangles, rhombi, parallelograms, and trapezoids.

For rectilinear figures in the plane, the concept of area is simple because everything reduces to a union of triangles. However, the course must deal with circles, and here limits must be used and the number π defined. The concept of limit can be employed intuitively without proofs. If the area or length of a circle is defined as the limit of approximating, inscribing, or circumscribing regular polygons, then π is either the area of the unit disk or the ratio of circumference to diameter, and heuristic arguments (see the glossary) for the equivalence of these two definitions would be given.

The concept of volume, in contrast with that of area, is not simple even for polyhedra and should be touched on only lightly and intuitively. However, the formulas for volumes and surface areas of prisms, pyramids, cylinders, cones, and spheres (Standard 9.0) should be memorized.

An important aspect of teaching three-dimensional geometry is to cultivate students' spatial intuition. Most students find spatial visualization difficult, which is all the more reason to make the teaching of this topic a high priority.

The basic mensuration formulas for area and volume are among the main applications of geometry. However, the Pythagorean theorem and the concept of similarity give rise to even more applications through the introduction of trigonometric functions. The basic trigonometric functions in the following standards should be presented in a geometry course:

> **18.0** Students know the definitions of the basic trigonometric functions defined by the angles of a right triangle. They also know and are able to use elementary relationships between them. For example, $\tan(x) = \sin(x)/\cos(x)$, $(\sin(x))^2 + (\cos(x))^2 = 1$.
>
> **19.0** Students use trigonometric functions to solve for an unknown length of a side of a right triangle, given an angle and a length of a side.

Finally, the Pythagorean theorem leads naturally to the introduction of rectangular coordinates and coordinate geometry in general. A significant portion of the curriculum can be devoted to the teaching of topics embodied in the next two standards:

17.0 Students prove theorems by using coordinate geometry, including the midpoint of a line segment, the distance formula, and various forms of equations of lines and circles.

22.0 Students know the effect of rigid motions on figures in the coordinate plane and space, including rotations, translations, and reflections.

The Connection Between Algebra and Geometry

These standards lead students to the next level of sophistication: an algebraic and transformation-oriented approach to geometry. Students begin to see how algebraic concepts add a new dimension to the understanding of geometry and, conversely, how geometry gives substance to algebra. Thus straight lines are no longer merely simple geometric objects; they are also the graphs of linear equations. Conversely, solving simultaneous linear equations now becomes finding the point of intersection of straight lines. Another example is the interpretation of the geometric concept of congruence in the Euclidean plane as a correspondence under an isometry of the coordinate plane. Concrete examples of isometries are studied: rotations, reflections, and translations. It is strongly suggested that the discussion be rounded off with the proof of the structure theorem: Every isometry of the coordinate plane is a translation or the composition of a translation and a rotation or the composition of a translation, a rotation, and a reflection.

Special attention should be given to the fact that a gap in Algebra I must be filled here. Standards 7.0 and 8.0 of Algebra I assert that:

1. The concept of slope of a straight line makes sense.
2. The graph of a linear equation is a straight line.
3. Two straight lines are perpendicular if and only if their slopes have a product of -1.

These facts should now be proved.

Additional Comments and Cautionary Notes

This section provides further comments and cautions in presenting the material in geometry courses.

Introduction to Proofs. An important point to make to students concerning proofs is that while the written proofs presented in class should serve as models for exposition, *they should in no way be a model of how proofs are discovered.* The perfection of the finished product can easily mislead students into thinking that they must likewise arrive at their proofs with the same apparent ease. Teachers need to make clear to their students that the actual thought process is usually full of false starts and that there are many zigzags between promising leads and dead ends. Only trial and error can lead to a correct proof.

This awareness of the nature of solving mathematical problems might lead to a deemphasis of the rigid requirements on the writing of two-column proofs in some classrooms.

Students' perceptions of proofs. The first part of the course sets the tone for students' perceptions of proofs. With this in

Chapter 3
Grade-Level
Considerations

Grades Eight Through Twelve

Geometry

Working on the level of axioms is actually more difficult for beginners than working with the theorems that come a little later in the logical development.

mind, it is advisable to discuss, mostly *without proofs*, those first consequences of the axioms that are needed for later work. A few proofs should be given for illustrative purposes; for example, the equality of vertical angles or the equality of the base angles of an isosceles triangle and its converse. There are two reasons for the recommendation to begin with only a few proofs. The foremost is that a complete logical development is neither possible nor desirable. This has to do with the intrinsic complexity of the structure of Euclidean geometry (see Greenberg 1993, 1–146). A second reason is the usual misconception that such elementary proofs are easy for beginners. Working on the level of axioms is actually more difficult for beginners than working with the theorems that come a little later in the logical development. This difficulty occurs because, on the one hand, working with axioms requires a heavy reliance on formal logic without recourse to intuition—in fact often *in spite of* one's intuition. On the other hand, working on the level of axioms does not usually have a clear direction or goal, and it is difficult to convince students to learn something without a clearly stated goal. If one so desires, students can always be made to go back to prove the elementary theorems *after* they have already developed a firm grasp of proof techniques.

Structured work with proofs. Students' first attempts at proofs need to be structured with care. At the beginning of the development of this skill, instead of asking students to do *many* trivial proofs after showing them the proofs of two or three easy theorems, it might be a good strategy to proceed as follows:

1. As early as possible, the students might be shown a generous number of proofs of substantive theorems so that they can gain an understanding of what a proof is before they write any proofs themselves.
2. As a prelude to constructing proofs themselves, the students might provide reasons for some of the steps in the sample (substantive) proofs instead of constructing extremely easy proofs on their own.
3. After an extended exposure to nontrivial proofs, students might be asked to give proofs of simple corollaries of substantive theorems.

The reason for steps 2 and 3 is to make students, from the beginning, associate proofs with real mathematics rather than perform a formal ritual. This goal can be accomplished with the use of *local axiomatics;* that is, if the proof of a theorem makes use of facts not previously proved, let these facts be stated clearly before the proof. These facts need not be previously proven but should ideally be sufficiently plausible even without a proof. Extensive use of local axiomatics would make possible, sufficiently early in the course, the presentation of interesting but perhaps advanced theorems. *In Appendix D, "Resource for Secondary School Teachers: Circumcenter, Orthocenter, and Centroid," the ideas in steps 2 and 3 are put to use to demonstrate how they might work.*

Development of geometric intuition. The following geometric constructions are recommended to develop students' geometric intuition. (In this context *construction* means "construction with straightedge and compass.") It is understood that all of them will be proved at some time during the course of study. The constructions that students should be able to do are:

- Bisecting an angle
- Constructing the perpendicular bisector of a line segment

- Constructing the perpendicular to a line from a point on the line and from a point not on the line
- Duplicating a given angle
- Constructing the parallel to a line through a point not on the line
- Constructing the circumcircle of a triangle
- Dividing a line segment into n equal parts
- Constructing the tangent to a circle from a point on the circle
- Constructing the tangents to a circle from a point not on the circle
- Locating the center of a given circle
- Constructing a regular n-gon on a given circle for $n = 3, 4, 5, 6$

Use of technology. This is the place to add a word about the use of technology. The availability of good computer software makes the accurate drawing of geometric figures far easier. Such software can enhance the experience of making the drawings in the constructions described previously. In addition, the ease of making accurate drawings encourages the formulation and exploration of geometric conjectures. For example, it is now easy to convince oneself that the intersections of adjacent angle trisectors of the angles of a triangle are most likely the vertices of an equilateral triangle (Morley's theorem). If students do have access to such software, the potential for a more intense mathematical encounter is certainly there. In encouraging students to use the technology, however, one should not lose sight of the fact that the excellent visual evidence thus provided must never be taken as a replacement for understanding. For example, software may give the following

heuristic evidence for why the sum of the angles of a triangle is 180°. When any three points on the screen are clicked, a triangle with these three points as vertices appears. When each angle is clicked again, three numbers will appear that give the angle measurement of each angle. When these numbers are added, 180° will be the answer. Furthermore, no matter the shape of the triangle, the result will always be the same.

While such exercises may boost one's belief in the validity of the theorem about the sum of the angles, it must be recognized that these angle measurements have added nothing to one's understanding of why this theorem is true. Furthermore, if one really wants to have a hands-on experience with angle measurements in order to check the validity of this theorem, the best way is to do it painstakingly by hand on paper. Morley's theorem, mentioned earlier, is another illustration of the same principle: evidence cannot replace proofs. The computer program would not reveal *the reason* the three points are always the vertices of an equilateral triangle.

Introduction to the coordinate plane. Students should know that the coordinate plane provides a concrete example that satisfies all the axioms of Euclidean geometry if the lines are defined as the graphs of linear equations $ax + by = c$, with at least one of a and b not equal to zero. Lines $a_1 x + b_1 y = c_1$ and $a_2 x + b_2 y = c_2$ are defined as parallel if (a_1, b_1) is proportional to (a_2, b_2), but (a_1, b_1, c_1) is not proportional to (a_2, b_2, c_2). The verification of the axioms is straightforward.

One should not lose sight of the fact that the excellent visual evidence thus provided by computers must never be taken as a replacement for understanding.

Algebra II

Algebra II expands on the mathematical content of Algebra I and geometry. There is no single unifying theme. Instead, many new concepts and techniques are introduced that will be basic to more advanced courses in mathematics and the sciences and useful in the workplace. In general terms the emphasis is on abstract thinking skills, the function concept, and the algebraic solution of problems in various content areas.

Absolute Value and Inequalities

The study of absolute value and inequalities is extended to include simultaneous linear systems; it paves the way for linear programming—the maximization or minimization of linear functions over regions defined by linear inequalities. The relevant standards are:

- **1.0** Students solve equations and inequalities involving absolute value.
- **2.0** Students solve systems of linear equations and inequalities (in two or three variables) by substitution, with graphs, or with matrices.

The concept of Gaussian elimination should be introduced for 2×2 matrices and simple 3×3 ones. The emphasis is on concreteness rather than on generality. Concrete applications of simultaneous linear equations and linear programming to problems in daily life should be brought out, but there is no need to emphasize linear programming at this stage. While it would be inadvisable to advocate the use of graphing calculators all the time, such calculators are helpful for graphing regions in connection with linear programming once the students are past the initial stage of learning.

Complex Numbers

At this point of students' mathematical development, knowledge of complex numbers is indispensable:

- **5.0** Students demonstrate knowledge of how real and complex numbers are related both arithmetically and graphically. In particular, they can plot complex numbers as points in the plane.
- **6.0** Students add, subtract, multiply, and divide complex numbers.

From the beginning it is important to stress the geometric aspect of complex numbers; for example, the addition of two complex numbers can be shown in terms of a parallelogram. And the key difference between real and complex numbers should be pointed out: The complex numbers cannot be linearly ordered in the same way as real numbers are (the real *line*).

Polynomials and Rational Expressions

The next general technique is the *formal* algebra of polynomials and rational expressions:

- **3.0** Students are adept at operations on polynomials, including long division.
- **4.0** Students factor polynomials representing the difference of squares, perfect square trinomials, and the sum and difference of two cubes.
- **7.0** Students add, subtract, multiply, divide, reduce, and evaluate rational expressions with monomial and polynomial denominators and simplify complicated rational expressions, including those with negative exponents in the denominator.

The importance of formal algebra is sometimes misunderstood. The argument against it is that it has insufficient real-world relevance and it leads easily to an overemphasis on mechanical drills. There seems also to be an argument for placing the study of exponential function ahead of polynomials in school mathematics because exponential functions appear in many real-world situations (compound interest, for example). There is a need to affirm the primacy of polynomials in high school mathematics and the importance of formal algebra. The potential for abuse in Standard 3.0 is all too obvious, but such abuse would be realized only if the important ideas implicit in it are not brought out. These ideas all center on the abstraction and hence on the generality of the formal algebraic operations on polynomials. Thus the division algorithm (long division) leads to the understanding of the roots and factorization of polynomials. The factor theorem, which states that $(x-a)$ divides a polynomial $p(x)$ if and only if $p(a) = 0$, should be proved; and students should know the proof. The rational root theorem could be proved too, but only if there is enough time to explain it carefully; otherwise, many students would be misled into thinking that *all* the roots of a polynomial with integer coefficients are determined by the divisibility properties of the first and last coefficients.

It would be natural to first prove the division algorithm and the factor theorem for polynomials with real coefficients. But it would be vitally important to revisit both and to point out that the same proofs work, verbatim, for polynomials with *complex* coefficients. This procedure not only provides a good exercise on complex numbers but also nicely illustrates the built-in generality of formal algebra.

Two remarks about Standard 7.0 are relevant: (1) a rational expression should be treated formally, and its function-theoretic aspects (the domain of definition, for example) need not be emphasized at this juncture; and (2) fractional exponents of polynomials and rational expressions should be carefully discussed here.

Quadratic Functions

The first high point of the course is the study of quadratic (polynomial) functions:

- **8.0** Students solve and graph quadratic equations by factoring, completing the square, or using the quadratic formula. Students apply these techniques in solving word problems. They also solve quadratic equations in the complex number system.
- **9.0** Students demonstrate and explain the effect that changing a coefficient has on the graph of quadratic functions; that is, students can determine how the graph of a

> The division algorithm (long division) leads to the understanding of the roots and factorization of polynomials.

parabola changes as a, b, and c vary in the equation $y = a(x-b)^2 + c$.

10.0 Students graph quadratic functions and determine the maxima, mimima, and zeros of the function.

What distinguishes Standard 8.0 from the same topic in Algebra I is the newly acquired generality of the quadratic formula: It now solves all equations $ax^2 + bx + c = 0$ with real a, b, and c regardless of whether or not $b^2 - 4ac < 0$, and it does so even when a, b, and c are *complex* numbers. Again it should be stressed that the purely *formal* derivation of the quadratic formula makes it valid for any object a, b, and c as long as the usual arithmetic operations on numbers can be applied to them. In particular, it makes no difference whether the numbers are real or complex. This premise illustrates the built-in generality of formal algebra. Students need to know every aspect of the proof of the quadratic formula. They should also be made aware (1) that with the availability of complex numbers, any quadratic polynomial $ax^2 + bx + c = 0$ with real or complex a, b, and c can be factored into a product of two linear polynomials with complex coefficients; (2) that c is the product of the roots and $-b$ is their sum; and (3) that if a, b, and c are real and the roots are complex, then the roots are a conjugate pair.

Standard 9.0 brings the study of quadratic polynomials to a new level by regarding them as a function. This new point of view leads to the exact location of the maximum, minimum, and zeros of this function by use of the quadratic formula (or, more precisely, by completing the square) without recourse to calculus. The practical applications of these results are as important as the theory.

Another application of completing the square is given in Standard 17.0, through which students learn, among other things, how to bring a quadratic polynomial in x and y without an xy term to standard form and recognize whether it represents an ellipse or a hyperbola.

Logarithms

A second high point of Algebra II is the introduction of two of the basic functions in all of mathematics: e^x and $\log x$.

11.0 Students prove simple laws of logarithms.

11.1. Students understand the inverse relationship between exponents and logarithms and use this relationship to solve problems involving logarithms and exponents.

11.2. Students judge the validity of an argument according to whether the properties of real numbers, exponents, and logarithms have been applied correctly at each step.

12.0 Students know the laws of fractional exponents, understand exponential functions, and use these functions in problems involving exponential growth and decay.

15.0 Students determine whether a specific algebraic statement involving rational expressions, radical expressions, or logarithmic or exponential functions is sometimes true, always true, or never true.

The theory should be done carefully, and students are responsible for the proofs of the laws of exponents for a^m where m is a rational number and of the basic properties of $\log_a x$: $\log_a (x_1 x_2) = \log_a (x_1) + \log_a (x_2)$, $\log_a (1/x) = -\log_a x$, and $\log_a(x^r) = r$

Students need to know every aspect of the proof of the quadratic formula.

$\log_a x$, where r is a rational number (Standard 15.0). The functional relationships $\log_a(a^x) = x$ and $a^{\log(t)} = t$ where a is the base of the log function in the second equation should be taught without a detailed discussion of inverse functions in general, as students are probably not ready for it yet. Practical applications of this topic to growth and decay problems are legion.

Arithmetic and Geometric Series

A third high point of Algebra II is the study of arithmetic and geometric series:

23.0 Students derive the summation formulas for arithmetic series and for both finite and infinite geometric series.

The geometric series, finite and infinite, is of great importance in mathematics and the sciences, physical and social. Students should be able to recognize this series under all its guises and compute its sum with ease. In particular, they should know by heart the basic identity that underlies the theory of geometric series:

$$x^n - y^n = (x-y)(x^{n-1} + x^{n-2}y + \cdots + xy^{n-2} + y^{n-1})$$

This identity gives another example of the utility of formal algebra, and the identity is used in many other places as well (the differentiation of monomials, for example). It should be mentioned that while it is tempting to discuss the arithmetic and geometric series using the sigma notation

$$\sum_{i=1}^{n},$$

it would be advisable to resist this temptation so that the students are not overburdened.

Binomial Theorem

Students should learn the binomial theorem and how to use it:

20.0 Students know the binomial theorem and use it to expand binomial expressions that are raised to positive integer powers.

18.0 Students use fundamental counting principles to compute combinations and permutations.

19.0 Students use combinations and permutations to compute probabilities.

In this context the applications almost come automatically with the theory.

Finally, Standards 16.0 (geometry of conic sections), 24.0 (composition of functions and inverse functions), and 25.0 may be taken up if time permits.

Trigonometry

Trigonometry uses the techniques that students have previously learned from the study of algebra and geometry. The trigonometric functions studied are defined geometrically rather than in terms of algebraic equations, but one of the goals of this course is to acquaint students with a more algebraic viewpoint toward these functions.

Students should have a clear understanding that the definition of the trigonometric functions is made possible by the notion of similarity between triangles.

A basic difficulty confronting students is one of superabundance: There are six trigonometric functions and seemingly an infinite number of identities relating to them. The situation is actually very simple, however. Sine and cosine are by far the most important of the six functions. Students must be thoroughly familiar with their basic properties, including their graphs and the fact that they give the coordinates of every point on the unit circle (Standard 2.0). Moreover, three identities stand out above all others: $\sin^2 x + \cos^2 x = 1$ and the addition formulas of sine and cosine:

3.0 Students know the identity $\cos^2(x) + \sin^2(x) = 1$:

> **3.1.** Students prove that this identity is equivalent to the Pythagorean theorem (i.e., students can prove this identity by using the Pythagorean theorem and, conversely, they can prove the Pythagorean theorem as a consequence of this identity).
>
> **3.2.** Students prove other trigonometric identities and simplify others by using the identity $\cos^2(x) + \sin^2(x) = 1$. For example, students use this identity to prove that $\sec^2(x) = \tan^2(x) + 1$.

10.0 Students demonstrate an understanding of the addition formulas for sines and cosines and their proofs and can use those formulas to prove and/or simplify other trigonometric identities.

Students should know the proofs of these addition formulas. An acceptable approach is to use the fact that the distance between two points on the unit circle depends only on the angle between them. Thus, suppose that angles a and b satisfy $0 < a < b$, and let A and B be points on the unit circle making angles a and b with the positive x-axis. Then $A = (\cos a, \sin a)$, $B = (\cos b, \sin b)$, and the distance $d(A, B)$ from A to B satisfies the equation:

$$d(A, B)^2 = (\cos b - \cos a)^2 + (\sin b - \sin a)^2$$

On the other hand, the angle from A to B is $(b - a)$, so that the distance from the point $C = \cos(b - a), \sin(b - a)$ to $(1, 0)$

The definition of the trigonometric functions is made possible by the notion of similarity between triangles.

is also $d(A, B)$ because the angle from C to $(1, 0)$ is $(b - a)$ as well. Thus:

$$d(A, B)^2 = (\cos(b - a) - 1)^2 + \sin^2(b - a)$$

Equating the two gives the formula:

$$\cos(b - a) = \cos a \cos b + \sin a \sin b$$

From this formula both the sine and cosine addition formulas follow easily.

Students should also know the special cases of these addition formulas in the form of half-angle and double-angle formulas of sine and cosine (Standard 11.0). These are important in advanced courses, such as calculus. Moreover, the addition formulas make possible the rewriting of trigonometric sums of the form $A\sin(x) + B\cos(x)$ as $C\sin(x + D)$ for suitably chosen constants C and D, thereby showing that such a sum is basically a displaced sine function. This fact should be made known to students because it is important in the study of wave motions in physics and engineering.

Students should have a moderate amount of practice in deriving trigonometric identities, but identity proving is no longer a central topic.

Of the remaining four trigonometric functions, students should make a special effort to get to know tangent, its domain of definition $(-\pi/2, \pi/2)$, and its graph (Standard 5.0). The tangent function naturally arises because of the standard:

7.0 Students know that the tangent of the angle that a line makes with the x-axis is equal to the slope of the line.

Because trigonometric functions arose historically from computational needs in astronomy, their practical applications should be stressed (Standard 19.0). Among the most important are:

13.0 Students know the law of sines and the law of cosines and apply those laws to solve problems.

14.0 Students determine the area of a triangle, given one angle and the two adjacent sides.

These formulas have innumerable practical consequences.

Complex numbers can be expressed in polar forms with the help of trigonometric functions (Standard 17.0). The geometric interpretations of the multiplication and division of complex numbers in terms of the angle and modulus should be emphasized, especially for complex numbers on the unit circle. Mention should be made of the connection between the nth roots of 1 and the vertices of a regular n-gon inscribed in the unit circle:

18.0 Students know DeMoivre's theorem and can give nth roots of a complex number given in polar form.

Mathematical Analysis

This discipline combines many of the trigonometric, geometric, and algebraic techniques needed to prepare students for the study of calculus and other advanced courses. It also brings a measure of closure to some topics first brought up in earlier courses, such as Algebra II. The functional viewpoint is emphasized in this course.

Mathematical Induction

The eight standards are fairly self-explanatory. However, some comments on four of them may be of value. The first is mathematical induction:

> **3.0** Students can give proofs of various formulas by using the technique of mathematical induction.

This basic technique was barely hinted at in Algebra II; but at this level, to understand why the technique works, students should be able to use the technique fluently and to learn enough about the natural numbers. They should also see examples of why the step to get the induction started and the induction step itself are both necessary. Among the applications of the technique, students should be able to prove the binomial theorem by induction and the formulas for the sum of squares and cubes of the first n integers.

Roots of Polynomials

Roots of polynomials were not studied in depth in Algebra II, and the key theorem about them was not mentioned:

> **4.0** Students know the statement of, and can apply, the fundamental theorem of algebra.

This theorem should not be proved here because the most natural proof requires mathematical techniques well beyond this level. However, there are "elementary" proofs that can be made accessible to some of the students. In a sense this theorem justifies the introduction of complex numbers. An application that should be mentioned and proved on the basis of the fundamental theorem of algebra is that for polynomials with real coefficients, complex roots come in conjugate pairs. Consequently, all polynomials with real coefficients can be written as the product of real quadratic polynomials. The quadratic formula should be reviewed from the standpoint of this theorem.

Conic Sections

The third area is conic sections (see Standard 5.0). Students learn not only the geometry of conic sections in detail (e.g., major and minor axes, asymptotes, and foci) but also the *equivalence* of the algebraic and geometric definitions (the latter refers to the definitions of the ellipse and hyperbola in terms of distances to the foci and the definition of the parabola in terms of distances to the focus and directrix). A knowledge of conic sections is important not only in mathematics but also in classical physics.

Limits

Finally, students are introduced to limits:

- **8.0** Students are familiar with the notion of the limit of a sequence and the limit of a function as the independent variable approaches a number or infinity. They determine whether certain sequences converge or diverge.

This standard is an introduction to calculus. The discussion should be intuitive and buttressed by much numerical data. The calculator is useful in helping students explore convergence and divergence and guess the limit of sequences. If desired, the precise definition of *limit* can be carefully explained; and students may even be made to memorize it, but it should not be emphasized. For example, students can be taught to prove why for linear functions $f(x)$, $\lim_{x \to a} f(x) = f(a)$ for any a, but it is more likely a ritual of manipulating ε's and δ's in a special situation than a real understanding of the concept. The time can probably be better spent on other proofs (e.g., mathematical induction).

Probability and Statistics

This discipline is an introduction to the study of probability, interpretation of data, and fundamental statistical problem solving. Mastery of this academic content will provide students with a solid foundation in probability and facility in processing statistical information.

Some of the topics addressed review material found in the standards for the earlier grades and reflect that this content should not disappear from the curriculum. These topics include the material with respect to measures of central tendency and the various display methods in common use, as stated in these standards:

6.0 Students know the definitions of the *mean, median,* and *mode* of a distribution of data and can compute each in particular situations.

8.0 Students organize and describe distributions of data by using a number of different methods, including frequency tables, histograms, standard line and bar graphs, stem-and-leaf displays, scatterplots, and box-and-whisker plots.

In the early grades students also receive an introduction to probability at a basic level. The next topic will expand on this base so that students can find probabilities for multiple discrete events in various combinations and sequences. The standards in Algebra II related to permutations and combinations and the fundamental counting principles are also reflective of the content in these standards:

1.0 Students know the definition of the notion of *independent events* and can use the rules for addition, multiplication, and complementation to solve for probabilities of particular events in finite sample spaces.

2.0 Students know the definition of *conditional probability* and use it to solve for probabilities in finite sample spaces.

3.0 Students demonstrate an understanding of the notion of *discrete random variables* by using them to solve for the probabilities of outcomes, such as the probability of the occurrence of five heads in 14 coin tosses.

The most substantial new material in this discipline is found in Standard 4.0:

4.0 Students are familiar with the standard distributions (normal, binomial, and exponential) and can use them to solve for events in problems in which the distribution belongs to those families.

Instruction typically flows from the counting principles for discrete binomial variables to the rules for elaborating probabilities in binomial distributions. The fact that these probabilities are simply the terms in a binomial expansion provides a strong link to Algebra II and the binomial theorem. From this base, basic probability topics can be expanded into the treatment of these standard distributions. In the binomial case students should now be able to define the probability for a range of possible outcomes for a set of events based on a single-event probability and thus to develop better understanding of probability and density functions.

The normal distribution, which is the limiting form of a binomial distribution, is typically introduced next. Students are not to be expected to integrate this distribution, but they can answer probability questions based on it by referring to tabled values. Students need to know that the mean and the standard deviation are parameters for this distribution. Therefore, it is important to understand variance, based on averaged squared deviation, as an index of variability and its importance in normal distributions, as stated in these standards:

5.0 Students determine the mean and the standard deviation of a normally distributed random variable.

7.0 Students compute the variance and the standard deviation of a distribution of data.

Standard 4.0 also includes exponential distributions with applications, for example, in lifetime of service and radioactive decay problems. Including this distribution acquaints students with probability calculations for other types of processes. Here, students learn that the distribution is defined by a scale parameter, and they learn simple probability computations based on this parameter.

4

Instructional Strategies

No single method of instruction is the best or most appropriate in all situations. Teachers have a wide choice of instructional strategies for any given lesson. Teachers might use, for example, direct instruction, investigation, classroom discussion and drill, small groups, individualized formats, and hands-on materials. Good teachers look for a fit between the material to be taught and strategies to teach it. They ask, What am I trying to teach? What purposes are served by different strategies and techniques? Who are my students? What do they already know? Which instructional techniques will work to move them to the next level of understanding? Drawing on their experience and judgment, teachers should determine the balance of instructional strategies most likely to promote high student achievement, given the mathematics to be taught and their students' needs.

The *Mathematics Content Standards* and this framework include a strong emphasis on computational and procedural competencies as a component of the overall goals for mathematics proficiency. The teaching of computational and procedural skills has its hazards, however. First, it is possible to teach computational and procedural skills in the absence of understanding. This possibility must be precluded in an effective mathematics program. A conceptual understanding of when the procedure should be used, what the function of that procedure is, and how the procedure manipulates mathematical information provides necessary constraints on the appropriate use of procedures and for detecting when procedural errors have been committed (Geary, Bow-Thomas, and Yao 1992; Ohlsson and Rees 1991).

Students gain a greater appreciation of the essence of mathematics if they are taught to apply those skills to the solution of problems. They may start to solve problems when armed with a small number of addition facts, and they need not wait to master all addition facts as a prerequisite to problem solving.

In a standards-based curriculum, good lessons are carefully developed and are designed to engage all members of the class in learning activities focused on student mastery of specific standards. Such lessons connect the standards to the basic question of why mathematical ideas are true and important. Central to the *Mathematics Content Standards* and this framework is the goal that all students will master all strands of the standards. Lessons will need to be designed so that students are constantly being exposed to new information while practicing skills and reinforcing their understanding of information introduced previously. The teaching of mathematics does not need to proceed in a strict linear order, requiring students to master each standard completely before being exposed to the next, but it should be carefully sequenced and organized to ensure that all standards are taught at some point and that prerequisite skills form the foundation for more advanced learning. Practice leading toward mastery can be embedded in new and challenging problems.

A particular challenge that the standards present to educators and publishers is the instruction of grade-level topics for students who have not yet mastered the expected content for earlier grades. One approach is to focus on the more important standards, as noted in Chapter 3, "Grade-Level Considerations." Bringing students up to grade-level expectations for those areas of emphasis will likely require (1) additional classroom time for mathematics, including time before, during, or after the instructional day; (2) the identification of the component skills that comprise each of the areas of emphasis; and (3) a reliable and valid means of assessing the degree to which individual students have mastered the component skills.

Instructional resources that help teachers to identify easily where these component skills were introduced in previous grades and that allow teachers to adapt these earlier-grade-level units into "refresher" lessons will be helpful. The development of such resources might require the development of a master guide on the organization of instructional units across grade levels. To achieve the goal of bringing students up to grade-level expectations requires that instructional materials be well integrated across grade levels. For example, instructional materials for fourth grade students should be written not only to address the fourth grade standards but also to prepare the foundation for mastery of later standards.

Organization of This Chapter

To guide educators in designing instructional strategies, this chapter is organized into three main sections:

1. "Instructional Models: Classroom Studies" provides an overview of research on student learning in classroom settings. In this section Table 1, "Three-Phase Instructional Model," provides a simple, research-based approach to instruction that all teachers may use.
2. "Instructional Models: View from Cognitive Psychology" provides a description of the research in cognitive psychology on the mechanisms involved in learning.
3. "General Suggestions for Teaching Mathematics" describes ways in which to organize the teaching of mathematics in kindergarten through grade twelve. Table 2, "Outline for Instruction of School-Based Mathematics," provides a convenient summary of the most important considerations for developing good lesson plans.

Instructional Models: Classroom Studies

Although the classroom teacher is ultimately responsible for delivering instruction, research on how students learn in classroom settings can provide useful information to both teachers and developers of instructional resources. This section provides an overview of student learning in classroom settings.

In conjunction with the development of this framework and the *Mathematics Content Standards,* the California State Board of Education contracted with the National Center to Improve the Tools of Educators, University of Oregon, in Eugene, to conduct a thorough review of high-quality experimental research in mathematics (Dixon et al. 1998). The principal goal of the study was to locate high-quality research about achievement in mathematics, review that research, and synthesize the findings to provide the basis for informed decisions about mathematics frameworks, content standards, and mathematics textbook adoptions.

From a total of 8,727 published studies of mathematics education in elementary and secondary schools, the research team identified 956 experimental studies. Of those, 110 were deemed high-quality research because they met tests of minimal construct and internal and external validity. The test of minimal construct looked at whether or not the study used quantitative measurements of mathematics achievement to report the effects of an instructional approach. To meet the internal validity criterion, the study had to use a true experimental design, have sufficient information to compute effect sizes, have equivalencies of groups at pretest, and use a representative and unbiased sample. External validity looked at whether or not the approach was implemented in settings representative of actual instructional conditions. The original report that the research team presented to the State Board of Education contained reviews of 77 of the qualifying studies; the most recent report includes information from all 110.

The reviewers cautioned readers about what their review did not do. Although a goal of the study was to find experimental support for the scope of instruction and the sequence of instructional topics, none of the high-quality experimental research studies addressed these important aspects of mathematics instruction. Instead, they looked only at findings relating to mathematics achievement. In addition, the review did not address such areas as improved attitudes toward mathematics or preferences for one mode of instruction over another.

Studies that met the high-quality review criteria indicated clear and positive gains in achievement from some types of instructional strategies. Perhaps most important, the review indicated marked differences in the effects of "conventional mathematics instruction" contrasted with interventions associated with high student achievement.

Two-Phase Model

As defined in the review, conventional mathematics instruction followed a two-phase model. In the first phase the teacher demonstrated a new concept, algorithm, or mathematical strategy while the students observed. In the second phase the students were expected to work independently to apply the new information, often completing work sheets, while the teacher might (or might not) monitor the students' work and provide feedback. This two-phase model, the researchers noted, was characterized by an abrupt shift in which students were expected "to know and independently apply the information newly taught moments earlier" (Dixon et al. 1998).

Three-Phase Model

More effective strategies may incorporate a variety of specific techniques, but they generally follow a clear three-phase pattern, as shown in Table 1, "Three-Phase Instructional Model."

The first phase. In the first phase the teacher introduces, demonstrates, or explains the new concept or strategy, asks questions, and checks for understanding. The students are actively involved in this phase instead of simply observing the teacher's lecture or demonstration. *Actively involved* should be thought of as a necessary but not sufficient characteristic of the first phase of effective mathematics instruction. (It is easy to imagine students actively involved in initial instructional activities that do not directly address the skills, concepts, knowledge, strategies, problem-solving competence, and understanding specified in the *Mathematics Content Standards.*) One way or another, all students must be involved actively in the introduction of new material. The corollary is that no student should be allowed to sit passively during the introduction of new material. The understanding demonstrated by a few students during this phase of instruction does not guarantee that all students understand. Teachers' classroom management and instructional techniques and the clarity and comprehensibility of initial instruction contribute to the involvement of all students. At the very least, active participation requires that the student attend to, think about, and respond to the information being presented or the topic being discussed.

The second phase. The second phase is an intermediate step designed to result in the independent application of the new concept or described strategy. This second step—the "help phase"—occurs when the students gradually make the transition from "teacher-regulation" to "self-regulation" (Belmont 1989). The details and specific instructional techniques of this phase vary considerably, depending on the level of student expertise and the type of material being taught. These techniques include any legitimate forms of prompting, cueing, or coaching that help students without making them dependent on pseudo-help crutches that do indeed help students but are not easily discarded. During this phase teachers also informally, but steadfastly, monitor student performance and move more slowly or more quickly toward students' independent, self-regulated achievement according to what the monitoring reveals about the students' progress.

Effective strategies may incorporate a variety of specific techniques, but they generally follow a clear three-phase pattern.

The third phase. In the third phase students work independently. In contrast with conventional lessons, however, the third phase is relatively brief instead of taking up most of the lesson time. This phase often serves in part as an assessment of the extent to which students understand what they are learning and how they will use their knowledge or skills in the larger scheme of mathematics.

This three-phase model is not rigid. If students do not perform well during the guided phase of instruction, then teachers should go back and provide additional clear and comprehensible instruction. If students do not perform well when they work independently, they should receive more guided practice and opportunities for application. And finally, if students perform well on a given topic independently but later display weaknesses with respect to that topic, then teachers should return to further guided instruction. This method is particularly critical when the topic at hand is clearly a prerequisite to further mathematics instruction and skill.

The table that follows shows each phase of the three-phase instructional model.

Table 1. Three-Phase Instructional Model

Phase 1	Phase 2	Phase 3
Teachers demonstrate, explain, question, and/or conduct discussions.	Teachers, individual peers, and/or groups of peers provide students with substantial help that is gradually reduced.	Teachers assess students' individual abilities to apply knowledge to new problems. Assessments can vary from informal to formal, depending on the immediate situation and goal of the particular lesson or lessons. The teacher should also review the goals of the lesson with students and tie goals back to the standard or standards.
Students are actively involved through answering questions, discussing topics, and/or attending to and thinking about the teacher's presentation.	Students receive feedback on their performance, correction, additional explanations, and other forms of assistance. Once the concept or procedure is understood, it is important for students to practice the material; otherwise, they are not likely to retain the just-learned information for long.	Students demonstrate their ability to work independently, generalize, and transfer their knowledge.

This three-phase model is framed by a beginning point (central focus) and by an ending point (closure), both of which are discussed next.

Central focus. In planning their lessons, teachers need to begin by identifying a central focus—the lesson's specific mathematical content and the goal of the lesson or sequence of lessons. Teachers also need to address the following concerns:

- The lesson or series of lessons should be focused on a clear instructional goal that is related to the mathematical content of the standards.
- The goal will typically be focused on fostering students' computational and procedural skills, conceptual understanding, mathematical reasoning, or some combination of these.
- The focus of a lesson or series of lessons is not simply to "cover" the required material but to build on previous knowledge and to prepare for future learning. Ultimately, the goals of any lesson are understood in the context of their relation to grade-level content, content covered in earlier grades, and content to be covered in later grades.

Closure. Closure of a lesson may take many forms. At the end of each lesson or series of lessons, students should not be left unsure of what has been settled and what remains to be determined. Whether the topic is covered within a single lesson or many, each lesson should contain closure that ties the mathematical results of the activities to the central goal of the lesson and to the goals of the overall series of lessons.

While current and confirmed research such as that reported in the Dixon study provides a solid basis on which to begin to design instruction, research from cognitive psychology provides insights into when and how children develop mathematical thinking.

Instructional Models: View from Cognitive Psychology

Initial competencies for natural abilities are built into the mind and brain of the child. These competencies develop during the child's natural social and play activities. Academic learning involves training the brain and mind to do what they were not designed by nature to do without help.

Natural Learning

The development of oral language is one example of natural learning. Young children naturally learn to speak as they listen to the speech around them. By the time they are five years old, they understand and can use approximately 6,000 to 15,000 words; they speak in coherent sentences using the basic conventions of the spoken language around them; and they can communicate effectively. Another example of natural learning is the early development of understanding about numbers. Starkey (1992) tested young children to determine their early understanding of arithmetic. He put up to three balls into a "search box" (a nontransparent box into which things are dropped and retrieved). At the age of twenty-four months (and sometimes younger), children would drop three balls into the box and then retrieve exactly three balls and stop looking for other balls in the box, showing that they were able to represent the number three mentally. They did this task without verbalizing, suggesting that a basic understanding of arithmetic is probably independent of language skills (Geary 1994, 41).

> Lessons should be focused on a clear instructional goal that is related to the mathematical content of the standards.

Certain features of geometry appear to have a natural foundation (Geary 1995). People know how to get from one place to another; that is, how to navigate in their environment. Being able to navigate and develop spatial representations, or cognitive maps, of familiar environments is a natural ability (e.g., picturing the location of the rooms in a house and the furniture in them). Without effort or even conscious thought, people automatically develop rough cognitive maps of the location of things in familiar environments, both small-scale environments, such as their house, and large-scale environments, such as a mental representation of the wider landscape (in three dimensions). Children's play, such as hide-and-seek, often involves spatial-related activities that allow children to learn about their environment without knowing they are doing so (Matthews 1992).

The brain and cognitive systems that allow us to navigate include an implicit understanding of basic Euclidean geometry. For example, we all implicitly know that the fastest way to get from one place to another is to "go as the crow flies"; that is, in a straight line. This is an example of natural conceptual knowledge. It is also sometimes taken as the first postulate of Euclidean geometry in school textbooks: A straight line can be drawn between any two points.

Academic Learning

The human brain and mind are biologically prepared for an understanding of language and basic numerical concepts. Without effort, children automatically learn the language they are exposed to, they develop a general sense of space and proportion, and they understand basic addition and subtraction with small numbers. Their natural social and play activities ensure that they get the types of experiences they need to acquire these fundamental skills. Not all cognitive abilities develop in this manner, however. In fact, most academic, or school-taught, skills do not develop in this manner because they are in a sense "unnatural" or formally learned skills (Geary 1995). As societies become more technically complex, success as an adult, especially in the workplace and also at home (e.g., managing one's money), involves more academic learning—skills that the brain and mind are not prewired to learn without effort. It is in these societies that academic schooling first emerged.

Schools organize the activities of children in such a way that they learn skills and knowledge that would not emerge as part of their natural social and play activities (Geary et al. 1998). If this were not the case, schooling would be unnecessary. But schooling is necessary, and it is important to understand why. Schooling is not necessary for the development of natural learning but is absolutely essential for academic learning. This is why teaching becomes so important. Teachers and instructional materials provide the organization and structure for students to develop academic skills, which include most academic domains; whereas nature provides for natural abilities. For academic domains this organization often requires explicit instruction and an explicit understanding of what the associated goals are and how to achieve them.

There are important differences in the source of the motivation for engaging in the activities that will foster the development of natural and academic abilities (Geary 1995). Children are biologically motivated to engage in activities, such as social discourse and play, that will automatically—without effort or conscious awareness—flesh out natural abilities, such as language. The motivation to engage in the activities that foster academic learning, in contrast, comes from the increasingly complex requirements of the larger society, not from the inherent interests of children. Natural play activities, or natural curiosity, of school-age students cannot be seen as sufficient means for acquiring academic abilities, such as reading, writing, and much of mathematics. The interests, likes, and dislikes of children are not a reliable guide to what is taught and how it is taught in school, although the interests of children probably can be used in some instructional activities. Once basic academic abilities are developed, natural interests can be used to motivate further engagement in some, but probably not all, academic activities.

Also relevant is intellectual curiosity, an important dimension of human personality (Goldberg 1992). People with a high degree of intellectual curiosity will seek out novel information and will often pursue academic learning on their own. Nevertheless, there are large individual differences in curiosity and, in fact, all other dimensions of personality. Some students will be highly curious and will actively seek to understand many things; others will show very little curiosity about much of anything; and most will be curious about some things and not others. If the goal is that all students meet or exceed specific content standards, then teachers cannot rely on natural curiosity to motivate all children to engage in academic learning.

In summary, natural mathematical abilities include the ability to determine automatically and quickly the number of items in sets of three to four items and a basic understanding of counting and very simple addition and subtraction; for example, that adding increases quantity (Geary 1995). These skills are evident in human infants and in many other species. Certain features of geometry, and perhaps statistics, also appear to have a natural foundation, although indirectly (Brase, Cosmides, and Tooby 1998).

Much of the content described in the *Mathematics Content Standards* is, however, academic. Mastering this content is essential for full participation in our technologically complex society; but students are not biologically prepared to learn much of this material on their own, nor will all of them be inherently motivated to learn it. That is why explicit and rigorous standards, effective teaching, and well-developed instructional materials are so important. The *Mathematics Content Standards,* teachers, and well-designed textbooks must provide for students' mathematical learning; that is, an understanding of the goals of mathematics, its uses, and the associated procedural and conceptual competencies.

General Suggestions for Teaching Mathematics

A general outline for approaching the instruction of school-based mathematics is presented in Table 2, "Outline for Instruction of School-Based Mathematics." Here, the teaching of mathematical units is focused on fostering the student's understanding of the goals of the unit and the usefulness of the associated competencies and on fostering general procedural and conceptual competence.

It is important to tell students the short-term goals and sketch the long-term implications of the mathematics they are expected to learn and the contexts within which the associated competencies, when developed, can be used. The short-term goals usually reflect the goal for solving a

Students are not biologically prepared to learn much of this material on their own, nor will all of them be inherently motivated to learn it.

Table 2. Outline for Instruction of School-Based Mathematics

Stating Goals and Uses	1.	Explicitly state the goal, or end-point, when a topic is first introduced.
	2.	State the immediate goal (e.g., the goal for this type of problem is to determine which pairs of numbers satisfy the equation) and, as appropriate, the long-term goal (e.g., this skill is used in many different types of applications, including . . . and is an essential part of mathematics because . . .).
Teaching Procedures	1.	Provide practice until the procedure is automatic; that is, until the student can use it without having to think about it. Automaticity will often require practice extending over many school years.
	2.	Provide practice in small doses (e.g., 20 minutes per day) over an extended period of time; practice on a variety of problem types mixed together.
	3.	Once automaticity is reached, include some additional practice of the procedure as part of review segments for more complex material. This practice form of review facilitates the long-term retention of the procedure.
Teaching Concepts	1.	When possible, present the material (e.g., word problems) in contexts that are meaningful to the student.
	2.	Solve some problems in more than one way. This exercise would typically be done after students have developed competence in dealing with the problem type.
	3.	Discuss errors in problem solving; use errors to diagnose and correct conceptual misunderstandings.

particular class of problem. For example, one goal of simple addition is to "find out the sum of two groups when they are put together." Knowing the goal of problem solving appears to facilitate the development of procedural and conceptual problem-solving competencies (Siegler and Crowley 1994).

Students should also be told some of the longer-term goals of what they are learning. This might include (1) stating what the students will be able to do at the end of the unit, semester, or academic year in relation to the mathematics standards; and (2) clarifying how current learning relates to the mathematics students will learn in subsequent years. It is also helpful to point out some of the practical uses of the new skills and knowledge being learned by linking them to careers and personal situations. *Studies of high school students indicate that making the utility of mathematics clear increases the student's investment in mathematical learning; that is, increases the number of mathematics courses taken in high school* (Fennema et al. 1981).

The usefulness of newly developing competencies might also be illustrated by having students use their skills in real-world simulations or projects; for example, figuring out how much four items cost at a store or using measurement and geometry to design a tree house. Assigning projects would not be the usual route to developing these competencies but would be a means of demonstrating their usefulness and providing practice. Projects might be used to introduce a difficult concept or to engage students in the unit. Using projects to stimulate interest and involvement must be weighed against the time they require and the extent of the mathematics learning. *Long projects with limited mathematical content and learning should be avoided, as should the use of brand names and logos.*

Procedural and Conceptual Competencies in Mathematics

Chapter 1, "Guiding Principles and Key Components of an Effective Mathematics Program," notes that the development of mathematical proficiency requires both procedural skills and conceptual knowledge and that these two components of mathematical competency are interrelated. It is now understood that the same activities—such as solving problems—can foster the acquisition of procedural skills and conceptual knowledge and can lead to the use of increasingly sophisticated problem-solving strategies (Siegler and Stern 1998; Sophian 1997). At the same time research in cognitive psychology suggests that different types of instructional activities will favor the development of procedural competencies more than conceptual knowledge, and other types of instructional activities will favor the development of conceptual knowledge more than procedural competencies (Cooper and Sweller 1987; Sweller, Mawer, and Ward 1983; Geary 1994). (See figure 1, "The Components of Conceptual, Procedural, and Reasoning Skills.")

Fostering procedural competencies. The learning of mathematical procedures, or algorithms, is a long, often tedious process (Cooper and Sweller 1987). To remember mathematical procedures, students must practice using them. Students should also practice using the procedure on all the different types of problems for which the procedure is typically used. Practice, however, is not simply solving the same problem or type of problem over and over again. Practice should be provided in small doses (about 20 minutes per day) and should include a variety of problems (Cooper 1989).

These arguments are based on studies of human memory and learning that indicate that most of the learning occurs during the early phases of a particular practice session (e.g., Delaney et al. 1998). In other words, for any single practice session, 60 minutes of practice is not three times as beneficial as 20 minutes. In fact, 60 minutes of practice over three nights is much more beneficial than 60 minutes of practice in a single night.

Moreover, it is important that the students not simply solve one type of problem over and over again as part of a single practice session (e.g., simple subtraction problems, such as $6 - 3$, $7 - 2$). This type of practice seems to produce only a rote use of the associated procedure. One result is that when students attempt to solve a somewhat different type of problem, they tend to use, in a rote manner, the procedure they have practiced the most, whether or not it is applicable. For example, one of the most common mistakes young students make in subtraction is to subtract the smaller number from the larger number

Practice, however, is not simply solving the same problem or type of problem over and over again.

Figure 1. The Components of Conceptual, Procedural, and Reasoning Skills

Goals to Be Achieved

Conceptual Competence	**Procedural Competence**	**Mathematical Reasoning Competence**
Knowing **what** to do	Knowing **how** to do it	Knowing **where** and **when** to do it
An implicit understanding of how to achieve the goal (Constraints are placed on the types of procedures used to achieve the goal.)	Behaviors that act on the environment to actually achieve the goal	An implicit understanding of the contexts within which conceptual and procedural competencies can be expressed

regardless of the position of the numbers. The problem shown below illustrates this type of error, which will be familiar to most elementary school teachers:

$$\begin{array}{r} 42 \\ -7 \\ \hline 45 \end{array}$$

Practicing the solving of simple subtraction problems (e.g., 7 − 2) is important in and of itself. Unthinking or rote application of the procedure is not the only cause of this type of error. In fact, this type of error should be a red flag for the teacher because it probably reflects the student's failure to understand regrouping.

One way to reduce the frequency of such procedural errors is to have the students practice problems that include items requiring different types of procedures (e.g., mixing subtraction and addition problems and, if appropriate, simple and complex problems). This type of practice provides students with an opportunity to understand better how different procedures work by making them think about which is the most appropriate procedure for solving each problem.

Ultimately, students should be able to use the procedure automatically on problems for which the procedure is appropriate. *Automatically* means that the procedure is used quickly and without errors and without the students having to think about what to do. Extensive practice, distributed over many sessions across many months or even years, might be needed for students to achieve automaticity for some types of mathematical algorithms.

Research indicates that long-term (over the life span) retention of mathematical competencies (and competencies in other areas) requires frequent refreshers (i.e., overviews and practice) at different points in the students' mathematical instruction (Bahrick and Hall 1991). One way to provide such a refresher is through a brief overview of related competencies when the students are moving on to a more complex topic.

In general, refreshers should focus on those basic or component skills needed to successfully solve new types of problems. For example, skill at identifying equal fractions, along with a conceptual understanding of fractions, will make learning to reduce fractions to their lowest terms much easier. These refreshers will provide the distributed practice necessary to ensure the automatic use of procedures for many years after the students have left school.

Fostering conceptual competencies. Fostering students' conceptual understanding of a problem or class of problems is just as important as developing students' computational and procedural competencies. Without conceptual understanding, students often use procedures incorrectly. More specifically, they tend to use procedures that work for some problems on problems for which the procedures are inappropriate.

Conceptual competency has been achieved when students understand the basic rules or principles that underlie the items in the mathematics unit. Students with this level of competency no longer solve problems according to the superficial features of the problem but by understanding the underlying principles. Students with a good conceptual understanding of the material are more flexible in their problem-solving approaches, see similarities across problems that involve the same rule or principle, make fewer procedural errors, and can use these principles to solve novel problems.

A number of teaching techniques can be used to foster students' conceptual understanding of problems (Cooper and Sweller 1987; Sweller, Mawer, and Ward 1983):

- First, when possible, the teacher should try to illustrate the problem by using contexts that are familiar and meaningful to students. In addition to fostering the students' conceptual understanding, familiar contexts will help students to remember what has been presented in class. Word problems, for example, should be presented in such contexts as home, school, sports, or careers.
- Second, after the students have developed some skill in solving this type of problem, the teacher should present a few problems that are good examples of the type of problem being covered and have the students solve the problems in a variety of ways. Psychological studies have shown that solving a few problems in many different ways is much more effective in fostering conceptual understanding of the problem type than is solving multiple problems in the same way. Solving problems in different ways can be done either as individual assignments or by the class as a whole. In the latter case, different students or groups of students might present suggestions for solving the problem. The students should be encouraged to explain why different methods work and to identify some of the similarities and differences among them.
- Third, for some problems, the teacher might have to teach one approach explicitly and then challenge the class to think of another way to solve the problem. Another approach is to have a student explain to the teacher why the teacher used a certain method to solve a problem (Siegler 1995). Having students explain what

> Without conceptual understanding, students often use procedures incorrectly.

someone else was thinking when he or she solved a problem facilitates their conceptual understanding of the problem and promotes the use of more sophisticated procedures during problem solving.

Errors should not simply be considered mistakes to be corrected but an opportunity to understand how the student understands the problem. Extensive studies of mathematical problem-solving errors indicate that most are not trivial but are systematic (VanLehn 1990). Generally, errors result from confusing the problem at hand with related problems, as in stating $3 + 4 = 12$ (confusing multiplication and addition), or from a poor conceptual understanding of the problem.

Typically, errors will result from confusion of related topics, such as addition and multiplication, a common memory retrieval error even among adults (Geary 1994). For other problems the error will reflect a conceptual misunderstanding, such as confusing the rules for solving one type of problem with those for solving a related type. For example, when students are first learning to solve simple subtraction problems, they are asked to subtract the smaller number from the bigger number (e.g., $6 - 3$) so that all of the differences are positive. From these problems, many students form the habit of taking the smaller number away from the larger number when subtracting. This rule works with simple problems but is often inappropriately applied to more complex problems; for example, those with a negative difference, such as $3 - 6$, or those that require borrowing, as in $42 - 7$. (Appendix B provides a sample East Asian mathematics lesson that can be used in staff development activities to stimulate a more extensive discussion about different ways of teaching mathematics.)

Having students provide several different ways to solve a problem and then spending time focusing on conceptual errors that might occur during this process can take up a significant portion of a lesson, but this method of instruction is usually worth the time. Students can practice using the associated procedures as part of their homework, or practice can occur in school, with a focus on practice and problem solving occurring on alternate days. A good policy is to be sure that the students have a conceptual understanding of the problem before they are given extensive practice on the associated procedures.

One way to monitor conceptual development is to ask students periodically to explain their reasoning about a particular mathematical concept, procedure, or solution. These explanations may be given either verbally or in writing. They provide a window for viewing the development of the students' understanding. During the early stages, for example, a student may be able to demonstrate a rudimentary grasp of a mathematical concept. At an intermediate stage the student may be able to give an appropriate mathematical formula. Students at advanced stages may be able to present a formal proof. Young students will not always have at their command the correct mathematical vocabulary or symbols, but as they progress, they should be encouraged to use the appropriate mathematical language.

For example, in the following problem students are asked to find the sum of the first n consecutive odd numbers. First one sees a pattern and conjectures at the general formula. The pattern is:

$1 = 1 = 1^2$
$1 + 3 = 4 = 2^2$
$1 + 3 + 5 = 9 = 3^2$
$1 + 3 + 5 + 7 = 16 = 4^2$
and so forth

Young students might explain this pattern in a number of ways. An older student might describe it as a formula. Students at the level of Algebra II might prove it by mathematical induction. (This is just one example of a problem that can be used to assess a student's development of mathematical reasoning.)

Overview and General Teaching Scheme

Figure 2, "General Framework for Teaching a Mathematics Topic," presents a flowchart that might be useful in preparing mathematics lessons for standards-based instruction. It should not be interpreted as a strict prescription of how standards-based instruction should be approached but as an illustration of many possible sequences of events, although most lessons or series of lessons should attempt to address most of these issues.

First, it is important to introduce the goals and specific mathematics content to be covered, along with some discussion of the different ways in which this type of mathematics is useful. For some topics the mathematics will be directly used in many jobs and even at home (e.g., shopping for the best price), or it might be a building block for later topics.

The next step is to present a brief overview (perhaps one part of a lesson) of the component skills needed to solve the problems to be introduced (e.g., review counting for lessons on addition, review simple addition in preparation for lessons on more complicated addition, and so forth). As part of this review, homework assignments that provide practice in these basic component skills would be useful. In this way the students will receive the extended and distributed practice necessary for them to retain mathematical procedures over the long term. At the same time this refresher will provide continuity from one unit to the next.

It is probably better to teach the conceptual features of the topic before giving the students extensive practice on the associated procedures. This instruction might be provided in three steps, although the number of steps used and the order in which they are presented will vary from one topic to the next. The first step is to introduce the basic concepts (e.g., trading or base-10 knowledge) needed to understand the topic. The second step is to design several lessons that involve solving a few problems in multiple ways and analyzing errors—this step will occur after the students understand the basic concepts and have some competence in solving the class of problem. The third step is to present an overview of the conceptual features of the topic, focusing on those areas where errors were most frequent. Once the initial class discussions of the topic have begun, the students can begin homework assignments (or alternating class assignments) in which they practice solving problems. Portions of these homework assignments can also be used as part of the refresher material for later lessons.

Homework

Student achievement will not improve much without study beyond the classroom. Homework should begin in the primary grades and increase in complexity and duration as students progress through school. To be an effective tool, homework must be a productive extension of class work. Its purpose and connection to class work must be clear to the teacher, student, and parents. The effective use of homework comes into play in Phase 2 of the three-phase instructional model outlined in Table 1. If the teacher chooses to

Figure 2. General Framework for Teaching a Mathematics Topic

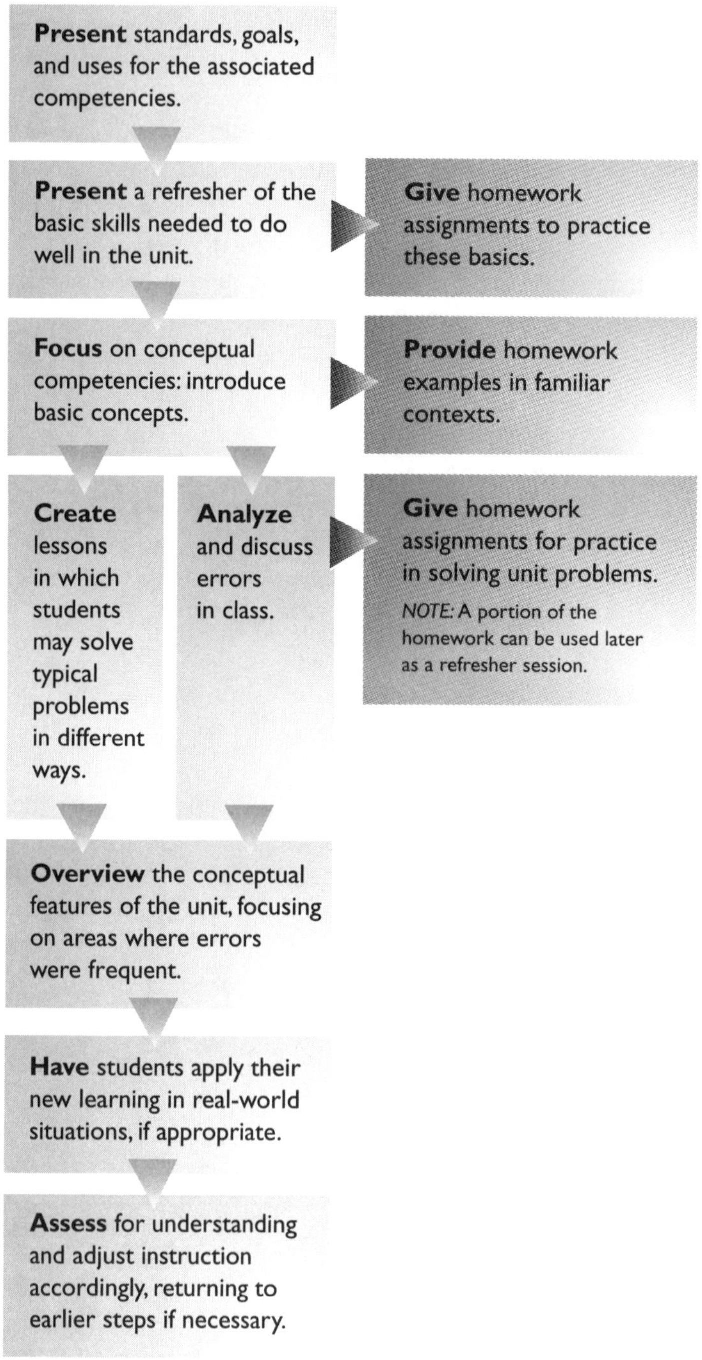

Present standards, goals, and uses for the associated competencies.

Present a refresher of the basic skills needed to do well in the unit.

Give homework assignments to practice these basics.

Focus on conceptual competencies: introduce basic concepts.

Provide homework examples in familiar contexts.

Create lessons in which students may solve typical problems in different ways.

Analyze and discuss errors in class.

Give homework assignments for practice in solving unit problems.

NOTE: A portion of the homework can be used later as a refresher session.

Overview the conceptual features of the unit, focusing on areas where errors were frequent.

Have students apply their new learning in real-world situations, if appropriate.

Assess for understanding and adjust instruction accordingly, returning to earlier steps if necessary.

allocate class time to discussion and feedback to students, he or she should ensure that this is productive instructional time for the class, a time when students are analyzing their errors and building their mathematical understanding. Instructional time is precious and should be used wisely. *Using substantial portions of the class period for homework is not an effective use of instructional time.* Using instructional time to review and correct common misconceptions evident from the teacher's analysis of the completed homework or using the last few minutes of a period to make sure that students understand the homework assignments, and how to complete them, can be effective uses of instructional time.

Several types of productive homework are outlined in Table 3.

Homework should increase in complexity and duration as students mature. Students studying for the *Advanced Placement* or *International Baccalaureate* examinations in mathematics will need additional study.

Table 3. Types of Homework

Exercises	Practice on skills is provided. It may involve practice problems similar to those worked in class or puzzles, games, or activities in which the skill is embedded. At advanced levels a single difficult problem can challenge students for an hour or more.
Lesson development	The student is involved in conceptual learning integral to classroom mathematics lessons.
Problem solving	Students get experience in solving problems and writing solutions. This activity is for individual work, but a student may use other persons as a resource.
Projects	Longer-term investigation and reporting in mathematics are provided, perhaps researching and describing an application of mathematics in the real world.
Study	Studying for understanding and preparing for exams are involved. This activity may be assigned or undertaken independently.

Source: Adapted from A. Holz. 1996. *Walking the Tightrope: Maintaining Balance for Student Achievement in Mathematics.* San Luis Obispo: California Polytechnic State University, Central Coast Mathematics Project.

Assessment

Inherent in the legislation that established the mathematics content standards is the explicit goal that every student will master or exceed world-class standards. The mathematics content standards set many learning goals that were previously viewed as being for only the most advanced students. Such ambitious goals demand a reexamination of the structures and assumptions that have driven the organization of kindergarten through grade eight mathematics programs and high school courses. To achieve world-class standards, each student must be continually challenged and given the opportunity to master increasingly complex and higher-level mathematical skills.

One problem associated with these goals is how best to detect and intervene with students who are at risk of falling behind or with those who can easily exceed grade-level standards. Optimally, no student should be allowed to slip behind for an entire semester or school year and, conversely, no student should be held back from progressing further just because the next level of learning is targeted for the next grade level.

Regular and accurate assessment of student progress in mastering grade-level standards will be essential to the success of any instructional program based on the mathematics content standards and this framework. Ideally, assessment and instruction are inextricably linked. The purposes of assessment that are the most crucial to achieving the standards are as follows:

- *Entry-level assessment.* Do students possess crucial prerequisite skills and knowledge? Do students already know some of the material that is to be taught?
- *Progress monitoring.* Are students progressing adequately toward achieving the standards?
- *Summative evaluation.* Have students achieved the goals defined by a given standard or a group of standards?

Taken together, these forms of assessment will provide a road map that leads students to mastery of the essential mathematical skills and knowledge described in the *Mathematics Content Standards*.

Entry-level assessment identifies what the student already knows and helps the teacher place the student at the most efficient starting point for his or her learning. A properly placed student will not waste time reviewing material he or she has already mastered. Nor will that student find himself or herself lost in instruction that is far beyond the student's current understanding.

Assessment that monitors student progress helps steer instruction in the right direction. It signals when alternative routes need to be taken or when the student needs to backtrack to gain more forward momentum.

Summative evaluation, which has characteristics similar to those of entry-level assessment, is done to determine whether the student has achieved at an acceptable level the goals defined in a standard or group of standards. Summative evaluation answers questions such as these: Does the student know and understand the material? Can he or she apply it? Has he or she reached a sufficiently high level of mastery to move on?

Similarities in Types of Assessments Across Grade Levels

All three types of assessment can guide instruction, and all three share critical characteristics across grade levels.

The exact purpose of each assessment item should be clear. Each item should be a reliable indicator of whether the student has the necessary prerequisite skills to move forward in mastering the standards. Some entry-level assessment items should measure mastery of the immediately preceding sets of standards. Others should measure the degree to which the student already has mastered some portion of what is to be learned next, if at all.

Entry-Level Assessment

Entry-level assessment needs to have a range and balance of items, some of which reach back to measure where students are, while others reach forward to identify those students who may already know the new material.

If entry-level assessments are used to compare the performance of students in the class or are used to establish a baseline for evaluating later growth, they must adhere to basic psychometric principles. That is, they must be:

1. Administered in the same conditions
2. Administered with the same directions
3. Scaled in increments small enough to detect growth

Progress Monitoring

In standards-based classrooms, progress monitoring becomes a crucial component of instruction for every student. It is only through such monitoring that teachers can continually adjust instruction so that all students are constantly progressing. No student should languish and be left behind because of a failure to recognize the need to provide him or her with extra help or a different approach. Similarly, students should not spend time practicing standards already mastered because of a failure to recognize that they need to move on.

In a sense everything students do during instruction is an opportunity for progress monitoring. Teachers should continually look for indicators among student responses and in student work. Monitoring can be as simple as checking for understanding or checking homework, or it may be a more formal type of assessment. Whatever form monitoring takes, it should occur regularly. In addition to regular monitoring to determine students' achievement of particular standards, more general monitoring should be done at least every six weeks.

Another form of monitoring is to make short, objective assessments to ensure that assessment of student learning is consistent for the entire class. Such measures must:

1. Use standardized administration procedures and tasks.
2. Document performance.
3. Be linked to items currently being taught.
4. Help teachers make instructional decisions and adjustments based on documented performance.
5. Indicate when direct interventions are needed for students who are struggling to master the standards.

The importance of using performance data as the basis for making well-informed adjustments to instruction cannot be overstated. Teachers need a solid basis for answering such questions as these:

- Should I move ahead or spend more time on the current phase of instruction?
- Are students able to practice what they have learned through independent activities, or do I need to provide additional instruction?
- Can I accelerate the planned instruction for some or all students and, if so, what is the best way to do that?

Summative Evaluation

Summative evaluation measures on a more formal basis the progress students have made toward meeting the standards. Typically, it comes at the end of a chapter or unit or school year. The most critical aspect of summative evaluation is that it measures the ability of students to transfer what they have learned to related applications. If one summative evaluation in the early grades is a test of computation, some or all of the problems should be new to the students; that is, problems that have not been used extensively during previous instruction.

This characteristic of summative evaluations addresses the concern many teachers have about "teaching to the test." Summative evaluations did not guide the development of the mathematics content standards; the standards provide the basis for developing summative evaluations. Further, summative evaluations are not mere reflections of retained knowledge but are the most valid and reliable indicator of depth of understanding.

Each of the three distinct types of assessment described in this chapter— entry-level assessment, progress monitoring, and summative evaluation—can help to guide effective instruction. Progress monitoring, in particular, can play a key role in developing and delivering curricula and instruction that lead to student achievement of the mathematics standards. Because this framework places substantial emphasis on integrating an assessment system with curricula and instruction, it is critically important that assessment and instruction be closely interrelated in ways that minimize any loss of instructional time while maximizing the potential of assessment to advance meaningful learning.

Special Considerations in Mathematics Assessment

A feature unique to mathematics instruction is that new skills are almost entirely built on previously learned skills. If students' understanding of the emphasis topics from previous years or courses is faulty, then it will generally be impossible for students to understand adequately any new topic that depends on those skills. For example, problems with the concept of large numbers as introduced in kindergarten and the first grade may well go unnoticed until the fifth grade, when they could cause students severe difficulty in understanding fractions. The biggest problem facing mathematics assessment is, therefore, how to devise comprehensive methods to detect the mastery of these basic learned skills.

There are many methods for assessment in mathematics, some of which will be mentioned in the next section. But certain methods, like timed tests, play a more basic role in mathematics assessment than they do in other areas of the curriculum in measuring understanding and skills and in checking whether students have an adequate knowledge base—whether they understand the material to the depth required for future success.

One of the key requirements for instructional materials discussed in Chapter 10 is that the materials provide teachers with resources and suggestions for identifying the basic prerequisite skills needed for the current courses and assessment material and suggestions that will help the teachers measure those skills. It is also recommended that this material include suggestions on how best to handle the most common types of difficulties that students will have.

Methods of Assessment in the Mathematics Curriculum

Many methods of assessment are available for testing knowledge in mathematics. Recently, one of the most pervasive methods, timed tests, has been the subject of intense scrutiny. A timed test requires that a certain number of items be completed within a fixed time limit. The following statement from the 1989 National Council of Teachers of Mathematics (NCTM) standards illustrates some of the issues:

> Students differ in their perceptions and thinking styles. An assessment method that stresses only one kind of task or mode of response does not give an accurate indication of performance, nor does it allow students to show their individual capabilities. For example, a timed multiple-choice test that rewards the speedy recognition of a correct option can hamper the more thoughtful, reflective student, whereas unstructured problems can be difficult for students who have had little experience in

A feature unique to mathematics instruction is that new skills are almost entirely built on previously learned skills.

exploring or generating ideas. An exclusive reliance on a single type of assessment can frustrate students, diminish their self-confidence, and make them feel anxious about, or antagonistic toward, mathematics (NCTM 1989, 202).

There is certainly an element of truth in this statement and, as is also advocated in the same document, other methods of assessment besides timed tests are appropriate in mathematics instruction.

> Many assessment techniques are available, including multiple-choice, short-answer, discussion, or open-ended questions; structured or open-ended interviews; homework; projects; journals; essays; dramatizations; and class presentations. Among these techniques are those appropriate for students working in whole-class settings, in small groups, or individually. The mode of assessment can be written, oral, or computer-oriented (NCTM 1989, 192).

All of these techniques can provide the teacher and the student with valuable information about their knowledge of the subject. However, they also represent a serious misunderstanding of what mathematics is and what it means to *understand mathematical concepts. Assessment methods such as timed tests play an essential role in measuring understanding—especially for the basic topics, the ones that must be emphasized.* If students are not able to answer questions in these areas relatively quickly, then their understanding of these topics is too superficial, has not been adequately internalized, *and will not suffice as a basis for further development.* For example, universities generally take for granted that students can perform basic calculations to the point of automaticity.

Again, the unique aspect of mathematics that was discussed previously must be emphasized. Mastery of *almost all the material* at each level depends on mastery of *all* the basic material at all previous levels. This requirement does not allow for superficial understanding, and the most efficient and reliable method for distinguishing between these levels of understanding remains the timed test.

The level of knowledge of basic topics needed for students to advance further requires that the topics be mastered to the level of automaticity. Consequently, the best method for assessing the basic topics is timed tests.

Students who do not have extensive experience during the school year with standardized, timed tests will be at a marked disadvantage in taking these types of tests; for example, those from the Standardized Testing and Reporting (STAR) Program, *Scholastic Achievement Test* (*SAT*) Program, and *American College Testing* (*ACT*) Program.

Readiness for Algebra

The step from grade seven mathematics to the discipline of algebra, which is one of the largest in the curriculum, can be more difficult to bridge than the previous steps from one grade level to the next. Moreover, the current requirement that algebra be taught at the eighth grade, whereas it was previously taught at the ninth or even the tenth grade, makes this step even greater.

Algebra I is a gateway course. Without a strong background in the fundamentals of algebra, students will not succeed in more advanced mathematics courses such as calculus. Nor will they be able to enter many high-technology and high-paying fields after graduation from high school (Paglin and Rufolo 1990). *It is therefore essential that the readiness of all students to take eighth-grade algebra be assessed at the end of the seventh grade, using reliable and valid assessment measures.*

One purpose of a seventh-grade assessment, as described previously, is to determine the extent to which students are mastering prealgebraic concepts and procedures. Another is to identify those students who lack the foundational skills needed to succeed in eighth-grade algebra and need further instruction and time to master those skills. This additional instruction may be provided through tutoring, summer school, or an eighth-grade prealgebra course leading to algebra in the ninth grade. The needs for such additional instruction will vary among the students, and it follows that *proper assessment at this level is crucial.*

Those students who have mastered foundational skills, as indicated by good performance on the algebra readiness test, would take algebra in the eighth grade.

The algebra readiness test should assess students' understanding of numbers and arithmetic, including knowledge of prime numbers and factoring, the rules for operating on integers (e.g., order of operations and associative and communicative properties), exponents, and roots. A thorough grounding in fractions, decimals, and percents, and the ability to convert easily from one to the other, is the fundamental algebra readiness skill. Testing students' readiness for algebra implies that options will be required for instructional materials at grade eight to accommodate students who are not ready to take the algebra course.

Statewide Pupil Assessment System

A major component of California's statewide testing system is the Standardized Testing and Reporting (STAR) Program. For mathematics, STAR, along with the Assessment of Applied Academic Skills currently under development, is the statewide system for summative assessment.

Standardized Testing and Reporting Program

STAR consists of three parts: (1) a standardized norm-referenced test; (2) an augmentation test aligned with the mathematics content standards; and (3) a standardized, norm-referenced primary language assessment. Characteristics of the STAR Program are that it:

- Requires the assessment of all students in English with a test approved by the State Board of Education
- Assesses achievement in reading, spelling, written expression, and mathematics in grades two through eight and in reading, writing, mathematics, history–social science, and science in grades nine through eleven
- Requires testing of academic achievement in the primary language for limited-English-proficient students enrolled for fewer than 12 months (optional thereafter)
- Generates the results of testing for individual students and reports to the public the results for schools, school districts, counties, and the state
- Disaggregates the results by grade level as to English proficiency, gender, and economic disadvantage for reporting to the public
- Provides both norm-referenced and standards-based results

The State Board of Education has adopted performance levels to be used in reporting the results of the augmented test: advanced, proficient, and basic, with an additional level designated as below basic. The levels correspond with those used by the National Assessment of Educational Progress. The augmented test addresses all

A thorough grounding in fractions, decimals, and percents, and the ability to convert easily from one to the other, is the fundamental algebra readiness skill.

the categories of the mathematics content standards.

Additional Components

Several additional components of the Statewide Pupil Assessment System enacted into law were being developed during the preparation of this framework for publication. The components include the following:

- Development of performance standards that define levels of student performance at each grade level in each of the areas in which the content standards have been adopted. "Performance standards gauge the degree to which a student has met the content standards and the degree to which a school or school district has met the content standards" (*Education Code* Section 60603[h]).
- Assessment of applied academic skills, based on the content and performance standards, that "requires pupils to demonstrate their knowledge of, and ability to apply, academic knowledge and skills in order to solve problems and communicate" in grades four, five, eight, and ten. "It may include . . . writing an essay . . . , conducting an experiment, or constructing a diagram or model" (*Education Code* Section 60603[b]).
- Adoption of a test to measure the development of reading, speaking, and writing skills in English for students whose primary language is not English.

The *Golden State Examination* Program completes the assessment picture in California and provides a measure of student achievement in several academic subjects normally taken at the middle school and high school levels. Participation in this program is voluntary, and students who do well on the tests receive a special honors designation. Work is under way to align the *Golden State Examination* with the mathematics content standards in the four core areas.

Universal Access

The diversity of California students presents unique opportunities and significant challenges for instruction. Students come to California schools with a wide variety of skills, abilities, and interests and with varying proficiency in English and other languages. The wider the variation of the student population in each classroom, the more complex becomes the teacher's role in organizing instruction in mathematics and ensuring that each student has access to high-quality mathematics instruction appropriate to the student's current level of achievement. The ultimate goal of mathematics programs in California is to ensure universal access to high-quality curriculum and instruction so that all students can meet or exceed the state's mathematics content standards. To reach that goal, teachers need assistance in designing curriculum and instruction to meet the needs of different groups of students. Through careful planning for modifying their curriculum, instruction, grouping, assessment techniques, or other variables, teachers can be well prepared to adapt to the diversity in their classrooms.

> *What the student already knows in mathematics should form the basis for further learning and study.*

This chapter refers frequently to students with special needs, a primary concern in considering the issue of universal access. Students in this category, who are discussed in this chapter, are special education pupils, English learners, students at risk of failing mathematics, and advanced learners.

Procedures that may be useful in planning for universal access are as follows:

- Assess each student's understanding at the start of instruction and continue to do so frequently as instruction progresses, and use the results of *assessment* for student placement and program planning.
- *Diagnose* the nature and severity of a student's difficulty and intervene quickly when students have trouble with mathematics.
- Engage in careful organization of resources and instruction and *planning* for adapting to individual needs. Be prepared to employ a variety of good teaching strategies, depending on the situation.
- *Differentiate* curriculum or instruction or both, focusing on the mathematics content standards and the key concepts within the standards that students must understand to move on to the next grade level.
- Use *flexible grouping* strategies according to the students' needs and achievement and the instructional tasks presented. (Examples of these strategies are homogeneous, semi-homogeneous, heterogeneous, large group, and small group. Individual learning is also an option.)
- *Enlist help* from others, such as mathematics specialists, mathematicians, special education specialists, parents, aides, other teachers, community members, administrators, counselors, and diagnosticians, when necessary, and explore the use of technology or other instructional devices as a way to respond to students' individual needs.

Alignment of Instruction with Assessment

One of the first tasks required of a school district is to determine its students' current achievement levels in mathematics so that each student or group of students can be offered a structured mathematics program leading to the attainment of all the content standards. What the student already knows in mathematics should form the basis for further learning and study.

Assessment is the key to ensuring that all students are provided with mathematics instruction designed to help them progress at an appropriate pace from what they already know to higher levels of learning. Knowing which standards the students have mastered, teachers and administrators can better plan the instructional program for each student or for groups of students with similar needs. Regardless of whether students are grouped according to their level of achievement in mathematics or some other factor, such as grade level, proficiency in English, or disability, assessment can be used to determine (1) which mathematical skills and understanding the student has already acquired; and (2) what the student needs to learn next. Regrouping can occur either within the classroom or across classrooms to place together students who are working on attaining a common set of mathematics standards.

For a variety of reasons, assessment of special needs students often reveals gaps in their learning. These gaps can be discovered through assessment, and instruction can be designed to remediate specific weaknesses without slowing down the students' entire mathematics program.

Successful Diagnostic Teaching

Students who have trouble in mathematics are at risk of failing to meet the standards, becoming discouraged, and eventually dropping out of mathematics altogether. The teacher should try to determine the cause of the learning difficulties, which may develop for a variety of reasons. Often a student placed in a class or program lacks the foundational skills and understanding necessary to complete new assignments successfully. If a student does not speak fluent English, for example, language barriers may be the cause of depressed mathematics scores. Sometimes a student may have a persistent misunderstanding of mathematics, or the student may have practiced an error so much that it has become routine. These problems may affect his or her ability to understand and complete assignments. Confusing, inadequate, or inappropriate instructional resources or instruction might be a contributing factor. A teacher can use the results of assessment and classroom observations to determine which interventions should be tried in the classroom and whether to refer the students to a student study team or to seek assistance from specialists.

Most learning difficulties can be corrected with good diagnostic teaching that combines repetition of instruction, focus on the key skills and understanding, and practice. For some students modification of curriculum or instruction (or both) may be required to accommodate differences in communication modes, physical skills, or learning abilities. Occasionally, in spite of persistent and systematic assistance, a student will continue to slip further behind or will show gaps and inconsistencies in mathematical skills that cannot be easily explained or remedied. For this small group of students, special assistance is a must.

To plan appropriate strategies for helping students who are experiencing difficulty in mathematics, teachers should consider the degree of severity of the learning difficulties according to the three major groups described next (Kame'enui and Simmons 1998). Intervention strategies will differ depending on the degree of severity.

Benchmark Group

Students in the benchmark group are generally making good progress toward the standards but may be experiencing temporary or minor difficulties. Although these students' needs are not intensive, they must be addressed quickly to prevent the students from falling behind. Often, the teacher can reteach a concept in a different way to an individual or a group of students or schedule a study group to provide additional learning time. Occasionally, parents can be enlisted to reinforce learning at home. Ideally, instructional resources will be organized in ways that make it easy for parents to do so. Some students may need periodic individual assistance, help from special education specialists, or assistive devices to ensure that they can succeed in the regular classroom. Once a student has grasped the concept or procedure correctly, additional practice is usually helpful. Most students will experience the need for temporary assistance at some time in their mathematical careers, and all students should be encouraged to seek assistance whenever they need it.

Most learning difficulties can be corrected with good diagnostic teaching

Strategic Group

Students in the strategic group may be one to two standard deviations below the mean on standardized tests. However, their learning difficulties, which must be examined with systematic and, occasionally, intensive and concentrated care, can often be addressed by the regular classroom teacher with minimal assistance within the classroom environment. A child-study team might be called on to discuss appropriate support for the student. In addition to reteaching a concept, the teacher may wish to provide specific assignments over a period of time for students to complete with a peer or tutor or by themselves at home. Regular study groups meeting before or after school, in the evenings, or on weekends can provide an effective extension of the learning time. Some students may need to take two periods of mathematics a day to master difficult content. Others, such as special education students, may need special modifications of curriculum or instruction to enable them to participate successfully in a mainstream classroom.

Intensive Group

Students in the intensive group are seriously at risk of failing to meet the standards as indicated by their extremely and chronically low performance on one or more measures. The greater the number of measures and the lower the performance, the greater is the students' risk. In general, these students perform more than two standard deviations below the mean on standardized measures and should be referred to a student study team for a thorough discussion of options. A referral to special education may be advisable. If eligible for special education services, these students will be given an individualized education plan, which will describe the most appropriate program for the student. Often, specialized assistance will be available through the special education referral. Such assistance may include intensive intervention by a qualified specialist, tutoring, the services of a classroom assistant, specialized materials or equipment (or both), or modification of assessment procedures.

Planning for Special Needs Students

Experienced teachers develop a repertoire of good instructional strategies to be used in special situations or with specific groups of students. Many of these strategies can be explicitly taught or can be embedded in instructional materials. To establish successful instructional strategies for all students, the teacher should:

1. Establish a safe environment in which the students are encouraged to talk and to ask questions freely when they do not understand.
2. Use a wide variety of ways to explain a concept or assignment. When appropriate, the concept or assignment may be depicted in graphic or pictorial form, with manipulatives, or with real objects to accompany oral instruction and written instructions.
3. Provide assistance in the specific and general vocabulary to be used for each lesson prior to the lesson and use reinforcement or additional practice afterward. Instructional resources and instruction should be monitored for ambiguities or language that would be confusing, such as idioms.
4. Set up tutoring situations that offer additional assistance. Tutoring by a qualified teacher is optimal. Peer or cross-age tutoring should be so

designed that it does not detract from the instructional time of either the tutor or tutee, and it should be supervised.

5. Extend the learning time by establishing a longer school day, a double period of mathematics classes, weekend classes, and intersession or summer classes.
6. Enlist the help of parents at home when possible.
7. Establish special sessions to prepare students for unfamiliar testing situations.
8. Ask each student frequently to communicate his or her understanding of the concept, problem, or assignment. Students should be asked to verbalize or write down what they know, thereby providing immediate insight into their thinking and level of understanding.
9. Use a variety of ways to check frequently for understanding. When a student does not understand, analyze why. This analysis may involve breaking the problem into parts to determine exactly where the student became confused.
10. Allow students to demonstrate their understanding and abilities in a variety of ways while reinforcing modes of communication that are standard in the school curricula.

Sequential or Simultaneous Instruction for English Learners

English fluency and academic achievement in the core curriculum do not need to be achieved simultaneously but may be addressed sequentially provided that, over a reasonable period of time, English learners do not suffer academically, as measured under the federal standards established in *Castaneda v Pickard* 648 F2d 989 (5th Cir 1981). It may be appropriate in some contexts to develop literacy in English initially, even if that approach delays progress toward achieving the standards. The English learner ultimately will have to acquire the English proficiency comparable to that of the average native speaker of English and to recoup any academic deficits resulting from the extra time spent on English language development. Thus, a school may elect first to focus on the development of English language skills and later to provide compensatory and supplemental instruction in mathematics so that an English learner can make up for academic deficiencies.

Differentiation in Pacing and Complexity

Advanced students and those with learning difficulties in mathematics often require systematically planned differentiation strategies to ensure appropriately challenging curriculum and instruction. The strategies for modifying curriculum and instruction for special education or at-risk students are similar to those used for advanced learners and can be considered variations along four dimensions: pacing, depth, complexity, and novelty. Two dimensions will be discussed here, pacing and complexity. Many of these strategies are good for all students, not just for those with special needs.

Pacing

Pacing is perhaps the most commonly used strategy for differentiation. That is, the teacher slows down or speeds up instruction. This strategy can be simple, effective, and inexpensive for many students with special needs (Benbow and

> The strategies for modifying curriculum and instruction can be considered variations along four dimensions: pacing, depth, complexity, and novelty.

Stanley 1996; Geary 1994). An example of pacing for advanced learners is to collapse a year's course into six months by eliminating material the students already know (curriculum compacting). Or students may move on to the content standards for the next grade level (accelerating). For students whose achievement is below grade level in mathematics, an increase in instructional time may be appropriate.

Complexity

Modifying instruction by *complexity* requires more training and skill on the part of the teacher and instructional materials that lend themselves to such variations. For students experiencing difficulty in mathematics, teachers should focus on the key concepts within the standards and eliminate confusing activities or variables.

Research shows that advanced students benefit most from a combination of acceleration and enrichment (Shore et al. 1991). These modifications can be provided either within a class or by the regrouping of students across classes or grade levels.

Differentiation for special needs students is sometimes questioned by those who say that struggling students never progress to the more interesting or complex assignments. This concern is often used to move struggling students along or to involve them in complex assignments even though they have not mastered the basics they need to understand the assignments. This framework advocates a focus on essential concepts embedded in the standards and frequent assessment to ensure that students are not just "passed along" without the skills they will need to succeed in subsequent grades. Struggling students are expected to learn the key concepts well so that they develop a foundation on which further mathematical understanding can be built.

Accelerating the Learning of At-Risk Learners

When students begin to fall behind in their mastery of mathematics standards, immediate intervention is warranted. Interventions must combine practice in material not yet mastered with instruction in new skill areas. Students who are behind will find it a challenge to catch up with their peers and stay current with them as new topics are introduced. Yet the need for remediation cannot be allowed to exclude these students from instruction in new ideas. In a standards-based environment, students who are struggling to learn or master mathematics need the richest and most organized type of instruction. This strategy will be very effective for the benchmark and strategic groups of students described by Kame'enui.

Students who have fallen behind, or who are in danger of doing so, need more than the normal schedule of daily mathematics. Systems must be devised to provide these students with ongoing tutorials. Such extra help and practice may occur in extra periods of mathematics instruction during the school day or during special after-school or Saturday tutorials or both.

As students in the eighth grade are assigned to formal mathematics courses, such as Algebra I or a first-year integrated course, new systems must be devised that provide students with an opportunity to stop and start over when they reach a point at which their eventual failure of the course is obvious. Requiring a student with intensive learning problems to remain in a course for which he or she lacks the skills to master the major concepts, and thereby to pass the course, wastes student learning time. Course and semester structures and schedules for classes should be reexamined and new structures devised so that students enrolled in such essential courses as Algebra I can begin again midyear or sooner in a class and move ahead to complete the full course.

For example, at the end of the first quarter or semester, classes might be reorganized so that students who are hopelessly lost and failing could enter a new course focused on strengthening their basic skills and restarting those students through the course curriculum. With such a structure the students would need to complete the final portion of the course in summer school or to be enrolled in a longer block of time for algebra during the regular school year. In either event the student would be more likely to succeed than if he or she were to remain in the Algebra I course for the entire school year and receive failing grades or to drop the class and miss out on instruction in mathematics altogether for the balance of the school year. In the past many students were reassigned to a basic general or remedial-level class, thereby limiting their opportunity to complete Algebra I, a critical gateway course.

Grouping as an Aid to Instruction

Research shows that what students are taught has a far greater effect on their achievement than how they are grouped (Mosteller, Light, and Sachs 1996). The first focus of educators should always be on the quality of instruction; grouping is a secondary concern. This framework recommends that educators use common sense about grouping. Grouping is a tool and an aid to instruction, not an end in itself. As a tool it should be used flexibly to ensure that all students achieve the standards. Instructional objectives should always be based on the standards and generally suggest grouping strategies.

For example, a teacher may discover that some students are having trouble understanding and using the Pythagorean theorem. Without this understanding they will have serious difficulties in algebra or geometry. It is perfectly appropriate, even advisable, to group those students who do not understand a concept or skill, find time to reteach the concept or skill in a different way, and provide additional practice. At the same time those students might be participating with a more heterogeneous mix of students in other classroom activities and in an evening study group in which a variety of mathematics problems are discussed.

Teachers must rely on their experiences and judgment to determine when and how to incorporate grouping strategies into the classroom. To promote maximum learning when grouping students, educators must ensure that assessment is frequent, that high-quality instruction is always provided, and that the students are frequently moved into appropriate instructional groups according to their needs. (*Note:* A more extensive discussion of grouping appears in Loveless 1998.)

English learners of different ages and with different primary languages may be grouped in the same classroom according to their English-language proficiency. These students would be educated in a structured English immersion program during a temporary transition period when the students acquire a good working knowledge of English. Mathematics instruction can be provided simultaneously (i.e., as English is acquired), or it may be scheduled sequentially (i.e., after a good working knowledge of English has been acquired). Once the necessary level of English is attained, these students would be instructed in an English mainstream setting.

Research shows that what students are taught has a far greater effect on their achievement than how they are grouped

Research on Advanced Learners

Advanced learners, for purposes of this framework, are students who demonstrate or are capable of demonstrating performance in mathematics at a level significantly above the performance of their age group. They may include (1) students formally identified by a school district as gifted and talented pursuant to California *Education Code* Section 52200; and (2) other students who have not been formally identified as gifted and talented but who demonstrate the capacity for advanced performance in mathematics. In California it is up to each school district to set its own criteria for identifying gifted and talented students; the percentage of students so identified varies, and each district may choose whether to identify students as gifted on the basis of their ability in mathematics. The research studies cited in this framework use the term *gifted students,* which is defined in most areas outside California in a more standardized way, in accord with nationally normed tests of achievement or intelligence. In that context *gifted students* usually refers to the top few percent of students who score at the highest percentiles on the tests.

A recent research study (Shore et al. 1991, 281) examined whether evidence exists to support 101 common practices in gifted education and found that very few practices were supported by solid evidence. However, the study also found that a combination of acceleration (students move on to material above grade level) and enrichment (students study topics in more depth or complexity or study related topics not covered in the normal curriculum) is supported by the research and leads to improved achievement for gifted students.

How to group advanced learners has been controversial. In a longitudinal study of grouping arrangements for over 1,000 elementary-age students, it was found that gifted students receiving an enriched and accelerated curriculum delivered in special schools, special day classes, and pullout programs made statistically significant increases in achievement in language arts, mathematics, science, and history–social science in comparison with gifted students who did not receive special programming (Delcourt et al. 1994).

The only type of programming arrangement that did not result in statistically significant improvement in achievement was enrichment offered in the regular heterogeneously grouped classroom. The reason for the lack of success was that even with the best of intentions, teachers did not have enough time to deliver the advanced or enriched curriculum they had planned for gifted students. Because most gifted students in California are served in the regular heterogeneously grouped classroom, teachers must ensure that enrichment or acceleration does occur, as argued for persuasively in the study (Delcourt et al. 1994).

Standards-based education offers opportunities for students who have the motivation, interest, or ability (or all of these) in mathematics to excel. Several research studies have demonstrated the importance of setting high standards for all students, including gifted students. The content standards in mathematics have provided students with goals worth reaching and identify the point at which skills and knowledge should be mastered. The natural corollary is that when standards are mastered, students should either move on to standards at higher grade

---

Research studies have demonstrated the importance of setting high standards for all students, including gifted students.

levels or focus on unlearned material not covered by the standards.

If, in a standards-based mathematics program, continual assessments of student achievement are calibrated to provide accurate measures of the uppermost reaches of each student's level of mathematical knowledge, the results will show that some students demonstrate mastery of the standards being studied by their peers, and a few others demonstrate complete mastery of all the standards for that grade level. A practical, common-sense response to students with such advanced ability would be to begin instruction in the next year's standards and continue to support the students' learning of new material and skills. A more common approach would be to provide such students with enrichment and depth in studying the standards for their grade level. However, with such an approach, enrichment or extension must actually occur and lead the student to complex, technically sound applications. There is a danger that enrichment experiences may become activities for marking time with interesting problems that do not actually contribute to higher learning or new insight.

If programs allow advanced students to move forward in the curriculum, schools will begin once more to produce young people who can compete at the highest levels in technological, mathematical, and science-based postsecondary and graduate studies and in professional fields yet to be developed. To fail to provide mathematically talented students with continuous new learning and challenges is to fail to develop a precious national resource and, in the name of equity, to consign all students to reach only the same average expectations (Benbow and Stanley 1996).

Accelerating the learning of talented students requires the same careful, consistent, and continual assessment of their progress as is needed to serve and advance the learning of average and struggling students. It is what schools do with the information from such assessments that challenges the educational system to reexamine its past practices. Elementary schools might be organized, for example, not around grade-level teams but around teams that span at least three grade levels. Such grade-span teams could provide mathematics instruction through flexible teaching focused on specific levels of the standards regardless of the students' grade levels. For example, in a team with students in grades three, four, and five, a mathematically advanced fourth grader might work with a group of fifth graders. Or a group of advanced third graders might work with fourth graders in need of an opportunity to relearn third-grade skills in an accelerated program. Through the use of such a system, learning groups might be regularly reorganized according to the students' mastery of standards so that the team members would be constantly working to move the students in the group forward.

Care must be taken in the design of standards-based programs to avoid the errors of the past. *In a standards-based classroom, the design of instruction demands dynamic, carefully constructed, mathematically sound lessons devised by groups of teachers pooling their expertise in helping children to learn.* These teams must devise innovative methods for using regular assessments of student progress to assign students to instructional groups in which teaching is targeted to ensure each student's progress toward mastery of world-class standards. (For more information on advanced learners, see the references section at the end of this framework.)

Responsibilities of Teachers, Students, Parents, Administrators

For students to achieve the high levels of mathematical understanding promoted in this framework, all parties to the educational process—teachers, students, parents, and administrators—need to play an active part in the process. Student success is maximized when all four parties coordinate their efforts. Teachers must take the responsibility for implementing a curriculum that provides students with a rigorous and meaningful experience in learning mathematics; students must respond by expending the effort to do as well as they can; parents must support and monitor student work; and administrators must provide the leadership that is central to a schoolwide mathematics program in which all students experience success.

Each of these four groups is responsible for lending support to the other three in pursuit of the goal of good education. This interlinked system of support consists of, at least, the following components:

- Establishing clear goals that are focused on students' learning of the mathematical content standards
- Developing a centralized system at the school level for monitoring students' academic progress

- Forging community and state partnerships based on clearly defined roles and responsibilities of all stakeholders
- Establishing a clear and consistent system of communication between school staff members and community stakeholders (e.g., parents, staff from local educational agencies, and business leaders)
- Having the goal that all students will leave school proficient in all grade-level standards covered in that particular school and ready for instruction in the grade-level standards in their next school, for enrollment in college-level mathematics, or for workplace training

Responsibilities of Teachers

Students' success in mathematics depends on the teacher more than on any other factor. Teachers are responsible for teaching mathematics in a way that provides exciting, balanced, high-quality programs. Teachers must strive to create an environment that enhances the mathematical understanding of all students. Teachers should be thoroughly versed in the content of the mathematics curriculum and be able to use various instructional strategies to help all students learn. They should continually evaluate the effectiveness of their teaching strategies and the usefulness of tasks and assignments, making adjustments when necessary or appropriate. In addition, teachers are responsible for assessing and monitoring student progress regularly and for providing help, enrichment, or acceleration as needed.

The *California Standards for the Teaching Profession* provides a framework of six instructional areas that can be applied to mathematics and across the curriculum (California Commission on Teacher Credentialing 1997):

1. Engaging and supporting all students in learning
2. Creating and maintaining effective environments for student learning
3. Understanding and organizing subject matter for student learning
4. Planning instruction and designing learning experiences for all students
5. Assessing student learning
6. Developing as a professional educator

Together with others in the school community, teachers are responsible for establishing good working relationships with parents and involving them as much as possible in their children's mathematics education. They should inform parents about appropriate roles for parents, expectations for student work, and student progress. Teachers should also make clear to parents the knowledge and skills contained in the *Mathematics Content Standards* and the nature of state assessments that measure student achievement.

Teachers who continue their own education and professional development throughout their careers are more likely than others to be acquainted both with mathematical content and with new developments in mathematics education. Their own professional growth should be as important to them as the growth and learning of their students.

Teachers' acquisition of this ever-expanding knowledge of the *what* and *how* of comprehensive and balanced mathematics instruction must be supported at all levels: state, county, district, and school. School district governing boards, superintendents, central office administrators, principals, mentors, teacher leaders, university faculty, subject-matter networks, and professional organizations also play key roles in developing and maintaining teacher expertise.

Ideally, teachers and administrators will support each other in a cooperative relationship that has as its goal the highest possible achievement in mathematics for all students.

Teachers should notify administrators of any issues that require administrative intervention. These issues may range from student discipline and classroom management problems to more immediate practical needs, such as additional instructional materials, repair of classroom fixtures, reduction of noise that interferes with learning and instruction, or excessive interruptions.

Responsibilities of Students

All students need a solid foundation in mathematics and all are capable of learning challenging mathematical content, although individual differences in educational outcomes are inevitable. Students must recognize that learning and progressing in mathematics result from dedication and determination.

An essential starting point for students is to take their mathematics studies seriously by working to become proficient and by participating actively in classroom instruction. They must make a commitment to attend all scheduled classes, complete all homework assignments, and acquire a determination to resolve problems and difficulties. Ideally, students must support one another and must cooperate with their teachers. They must persist when the mathematics content becomes challenging.

Each level of mathematical growth requires mastery of the preceding level. At each stage students must learn and reinforce basic skills and also make a concerted effort to understand mathematical concepts and to apply those concepts and skills to problem solving. Students should also be encouraged to learn, understand, and master the many different dimensions of mathematics by learning to reason mathematically, to employ a variety of methods for problem solving, and to communicate and validate solutions mathematically, giving accurate and detailed proofs where needed. Although good teachers can provide encouragement and help, students are responsible for their own learning, and no one can learn for them.

Realistically, no one can expect all students to enter mathematics instruction with high levels of personal responsibility or motivation to learn mathematics. Students who are not highly responsible or motivated should not be ignored, however. Parents, the school, other stakeholders, and even the instructional materials should contribute to helping all students accept personal responsibility. Good schoolwide management plans, for example, can improve attendance dramatically. Many students need to be taught good study skills. Effective instruction will give students the success that, in turn, will motivate them to work toward more and greater success.

Teachers can provide needed support by stressing that mathematics learning requires considerable effort from all students and that persistent effort will greatly improve their learning and achievement. In other words gaining competency in mathematics does not require inherent mathematical talent, but it does require sustained effort and hard work (Stevenson et al. 1990). Finally, good in-class management systems will reduce the extent to which one student's lack of responsibility infringes on the efforts of others.

Responsibilities of Parents and Families

Ideally, all parents should be strong advocates of their children's education. Many parents may need assistance and encouragement from the school to support them in this role.

Whether students are underachieving, average, gifted, or in need of individual attention, parents should recognize their own and their children's role in learning mathematics and achieving optimal success. They should know the specific academic standards their children are to meet at each grade level, and they should be able to monitor their children's performance and provide extra help when needed. Parents should be responsible for obtaining information regarding their children's progress and know how to interpret that information appropriately. Above all, they should encourage a positive attitude toward mathematics.

Parents are their children's first teachers. A child's early experiences with mathematics at home can provide an important foundation for learning the content standards for kindergarten (Saxe, Guberman, and Gearhart 1987). Parents and other family members can nurture and stimulate mathematics development in their children and, for many children, will need to be involved in their children's mathematics program at all grade levels (Stevenson et al. 1990).

However, schools must take greater responsibility to support the early mathematics development of children who are less fortunate and do not benefit from an educated, supportive family environment. Such support may require after-school homework, transportation services to bring children to school early for extra tutoring, extended tutoring support, and similar kinds of programs.

Parents should be encouraged to reinforce school learning of mathematics at home by setting aside a place and the time for homework. They should check regularly with their children to make sure that the assignments have been completed and the material understood. It would be very helpful if, in addition, they were to participate in both school and districtwide activities that may affect their children's education, such as developing curricula, selecting instructional materials and assessment instruments, and establishing local educational goals.

Parents should feel comfortable requesting information from the school about the existing mathematics curriculum and the standards for content and performance. The school should encourage these requests and make such information readily available by providing convenient access to instructional and enrichment materials. Ideally, each textbook will be accompanied by a publisher's handbook to help parents monitor their children's lessons.

Frequently, parents are reluctant to become involved in mathematics education because of their own lack of confidence with mathematics. Although parents do not have to be mathematicians to participate actively in their children's mathematics program, schools should consider inviting parents to participate in professional development opportunities designed for the improvement of their personal mathematics proficiency. Parents can support their children's mathematics education by ensuring that their children complete homework assignments and regularly attend school and by ensuring that their children recognize the importance of achievement in mathematics.

Responsibilities of Administrators

Administrators are responsible for promoting the highest-quality mathematics programs. It is their job to hire and assign appropriately credentialed, skilled, and effective mathematics teachers and provide mentoring and professional support for teachers when necessary. They should also monitor teachers' implementation of the mathematics curriculum and evaluate teachers' ability to teach the curriculum and to assess student performance and progress. Administrators should ensure that the curriculum is coherent and consistent with the state standards and the guidelines in this framework. As much as possible, they should ensure continuity in the mathematics curriculum from classroom to classroom and among grade levels.

Administrators play a critically important leadership role. They are responsible for seeing that the schools maintain high standards for their mathematics offerings and that quality programs are implemented for all students. Effective administrative leadership encourages teachers, students, and parents to recognize the importance of a quality mathematics program and actively support its implementation. Administrators should respond appropriately to teachers' concerns about student learning. They are also responsible for ensuring that each student has adequate and appropriate instructional materials and supplies of suitable quality.

Any explicit recommendation on what administrators can do to promote a successful program in mathematics is likely to involve funding and therefore decrease its chance of implementation. There is, however, an urgent recommendation that can be implemented without funding: *Administrators must provide uninterrupted instruction time to teachers.* Interruptions, such as intercom announcements, call slips and messages for students, and nonmathematical activities, such as field trips, athletic events, and other extracurricular activities, that take place during class time have a significant negative impact on mathematics instruction. Administrators are responsible for ensuring that such interruptions are limited as much as possible to real emergencies, with class time consistently protected and considered inviolable.

Administrators should develop among staff, students, and parents a climate of partnership, teamwork, collaboration, and innovation to sustain high achievement. Also essential is that administrators provide opportunities for teachers to develop collegial and collaborative relationships within their grade levels, including allowing adequate time for meetings and informal discussions. More precisely, administrators can use the suggestions in the list that follows to develop an environment that encourages achievement. Administrators should:

- Allot time at staff meetings (and perhaps invite experts to visit the school) to discuss recent research articles in mathematics and their usefulness and application to current school practices and the mathematics improvement plan.
- Provide time for monthly grade-level meetings that focus on assessing student progress toward achieving the standards and on modifying programs to improve student progress.
- Provide time for teachers to visit other classrooms, both within the school and at model implementation sites, to observe and discuss instructional strategies and materials used by teachers

---

Administrators should ensure that the curriculum is coherent and consistent with the state standards and the guidelines in this framework.

who are highly successful in fostering student learning and achievement in mathematics.
- Provide time for professional development, including many opportunities for staff members to receive coaching from those with expertise in teaching mathematics.
- Provide, if necessary, the resources needed to hire specially trained mathematics teachers so that these teachers can provide support to the faculty members for the implementation of the standards.
- Communicate to parents the school district's expectations for student performance, including the content of the state standards and the nature of state assessments that measure student achievement.

Professional Development

The goals of professional development are to provide classroom teachers with the knowledge and skills they will need to implement the mathematics content standards and to ensure that prospective teachers will be prepared to teach mathematics effectively. Assistance in achieving these goals is provided through the topics addressed in this chapter: district programs for professional development, professional development and retention of new teachers, long-term professional development, issues to consider in designing professional development programs, issues to consider in implementing new curricula, and undergraduate preparation.

District Programs for Professional Development

Because the new standards are extremely rigorous, implementing them presents a challenge to teachers and students. While the standards articulate what students need to know and be able to do to compete in a global economy, most California students currently perform at unacceptably low levels. The challenge of implementing the standards will be daunting because (1) the significant number of students not currently performing at the levels expected by the standards means that instruction will need to be organized and presented efficiently and effectively; and (2) the increased rigor of the standards will require many teachers at all levels to become more knowledgeable about mathematics and ways in which to teach it.

Given the enormity of the challenge, teachers must learn the most effective techniques for teaching mathematics, such as those described in Chapter 4, "Instructional Strategies." Teachers must be prepared to address the needs of all their students. In some schools teachers will find that most students have a wide background and understanding of mathematics. In other schools teachers will encounter large numbers of students who have not learned foundational content. The professional development for teachers in the short term must be geared to the specific challenges that teachers will face in implementing the standards in their classrooms. Teachers need information based on what has been proven to work with students. Professional development should be based only on proven techniques and theories.

Ongoing professional development is expected of all teachers. Professional development for immediate classroom application should take place locally and regularly. Effective professional development is long term and focused on increasing teachers' mathematical knowledge and ability to teach the subject.

Professional Development and Retention of New Teachers

All too often, new teachers leave the profession during or just after their first year of teaching, thereby wasting much of the huge investment in their education. To help in solving this problem, school administrators and colleagues must take steps to help new teachers succeed in the classroom. Careful placement and active mentoring can help, as can the activities for all teachers discussed next. These activities can alleviate the isolation that can be a problem for all teachers but which is most acute during the first year of teaching.

To overcome the potential for isolation, teachers should be encouraged to set up ongoing collegial support, the focus of which is to share successful lessons and teaching approaches, and to coach one another in ways to improve student achievement. School and district administrators should support such in-house efforts by making time and space available and, when possible, by bringing in qualified mathematics specialists to help with these developmental activities.

All professional development programs that focus on mathematics should use this framework to address the needs of the teachers involved. Further, teachers should be active participants in planning and organizing meetings and attending short courses and workshops offered by local districts, colleges, universities, independent consultants, and professional organizations to ensure that the teachers' needs are being met. These programs must be

Effective professional development is long term and focused on increasing teachers' mathematical knowledge and ability to teach the subject.

assessed independently to determine their usefulness to teachers; for example by leading to improvement in student mathematics achievement.

Long-Term Professional Development

One goal of long-term professional development is to sustain and increase teachers' understanding of mathematics; that is, teachers' procedural competencies, conceptual knowledge, and ability to use these competencies and knowledge in problem-solving situations.

Long-term professional development must be expected, actively encouraged, and rewarded both by school administrators and by state and national efforts. Support from institutions of higher education and other institutions with sufficient expertise must be enlisted in an effort to make opportunities for high-quality, long-term professional development readily available to all mathematics teachers. Long-term professional development programs in mathematics should be routinely subject to external assessment to ensure that they achieve their goals toward enhancing the mathematical competencies and knowledge of teachers. Teachers should be encouraged to share the benefits of their long-term professional development, as appropriate, with their colleagues in local in-service training programs. Teachers' leadership and participation in national and local professional organizations that support the spirit and letter of this framework and the state standards are valued as a hallmark of the teachers' professionalism.

Since professional development is essential to implementing the high standards defined within this framework, a variety of issues should be considered when such activities are being designed. The list of issues that follows is neither a check-off list of what should be done nor an exhaustive compilation of items. Indeed, any particular focused and in-depth professional development program may be able to deal with only a few issues each year. In deciding how to balance these considerations, teachers and school district administrators need a clear understanding of their goals for professional development.

Issues to Consider in Designing Professional Development Programs

Persons planning activities or programs for professional development need to consider the following issues:

- The emphasis and focus of all professional development programs in mathematics is on the effective implementation of the state standards and the guidelines presented in this framework. Individuals who provide professional development programs must be willing and able to demonstrate the effectiveness of their recommendations for the typically diverse California classroom. *Those teaching mathematics to teachers must themselves be competent with mathematics and competent teachers of teachers.* Teachers of classroom management must be competent at managing classrooms effectively, and those helping teachers learn effective instructional strategies must be competent themselves at demonstrating any such strategies in classrooms.
- Programs with lasting influence are usually long term and locally based, with teachers playing a substantial role in planning and implementation. The

program should receive regular feedback from teachers and be modified accordingly.
- Effective professional development is needed to enable teachers to maximize instructional time. Teachers can be helped in resolving both mathematics-specific classroom management issues (e.g., organizing, distributing, or collecting concrete materials) and more generalized management concerns (e.g., dealing with an inappropriately high classroom noise level, frequent tardiness or absences, or inattention).
- Program activities should be structured to raise teachers' proficiency in mathematics. As their proficiency increases, teachers will find that their comfort levels in using their new mathematical knowledge will also increase.
- Program participants need to discuss and understand fully what it means to balance the learning of mathematical procedures, the understanding of mathematical concepts, and the ability to engage in mathematical problem solving across different grade and achievement levels (see Chapters 1 and 4).
- Programs need to help teachers and administrators, through in-service training and sharing, expand their understanding of student differences, diverse cultures, and specific instructional implications and accommodations.
- Teachers, while accommodating differences among students, must know which standards provide the core mathematics foundation for all students at each grade level (see Chapter 3, "Grade-Level Considerations").
- Parent involvement can substantially influence student success (Slavin, Karweit, and Wasik 1994). Programs should help teachers develop various strategies to help parents become effectively involved in the mathematics education of their children.
- Professional development programs should focus both on mathematics proficiency for students and on those instructional strategies that best achieve it (see Chapters 1 and 4).
- Teachers need to learn about various forms of assessment, including methods of conducting frequent assessments to monitor students' progress, along with remedies when such assessments reveal that students are not achieving at grade-level expectations.
- Teachers need time to discuss with their peers ways in which to implement concepts that were presented in professional development programs.

Issues to Consider in Implementing New Curricula

Many professional development issues need to be addressed with the implementation of the standards. Some of the more pressing issues are as follows:

- Professional development needs to provide teachers with a clear understanding of standards-based mathematical expectations. Students need to know the goals and uses of the mathematics they are taught (see Chapter 4), and teachers need to understand the basic goals of the standards and the importance of achieving those goals.
- Teachers need to understand how the grade-level content they are teaching is related to the content taught in previous grades and how their teaching will prepare students for the mathematics to be introduced in later grades.

Teachers need to understand how the grade-level content they are teaching is related to the content taught in previous grades and how their teaching will prepare students for the mathematics to be introduced in later grades.

- As described in Chapter 10, "Criteria for Evaluating Instructional Resources," well-designed instructional materials will greatly facilitate this goal. But, at the same time, in-service training or other activities will also be needed to show teachers how their teaching is an integral part of all grade-level standards and how they can develop strategies for linking their teaching to material for earlier and later grades (e.g., identifying review materials for improving their students' foundational skills).
- Phase-in strategies for new curricula must be considered carefully. To maintain momentum, teachers should be provided the necessary in-class support to implement new programs consistently and according to a given timeline.

Every student of mathematics deserves to be taught by a teacher who has both the mathematical knowledge and teaching skills needed to implement the standards at each student's achievement level. The teacher must present mathematics in ways that allow students to experience the excitement and joy of doing mathematics and to attain mathematics proficiency. Such teachers are a precious resource that must be nurtured at all levels, and can, as math specialists, assist students and teachers alike. Young adults who love mathematics must be recruited into the profession and supported through preservice preparation and in-service education. They must be helped in (1) learning mathematics; and (2) developing a repertoire of effective teaching strategies that allow them to implement a curriculum that balances problem solving, conceptual understanding, and procedural skills, as described in Chapters 1 and 4. California's teachers deserve and must receive the greatest possible support in this endeavor so that they can succeed in becoming skilled mathematics teachers.

Given the shortage of highly trained mathematics teachers in California, schools and undergraduate institutions must actively encourage talented mathematics students to go into teaching careers. Undergraduate internships in kindergarten through grade twelve classrooms, followed up with guided reflection and discussion, can be effective recruitment tools and can enhance the value of undergraduate mathematics education.

Undergraduate Preparation

Teachers need a background in mathematics that is considerably deeper and broader than the mathematics they are expected to teach. Teachers at lower grade levels need this background to understand how their teaching relates to the mathematics content in later grades. Teachers at each grade level need to understand what students will encounter in subsequent grades, because teachers will then know which foundational skills taught at their grade level deserve the greatest attention and emphasis. To achieve this understanding, teachers need to acquire a mathematical breadth that enables them to comprehend the interrelationships of mathematical concepts and procedures across strands. Teachers' mathematical depth should enable them to understand the dependence of mathematical ideas on one another; for example, adding rational expressions in algebra depends on adding fractions in arithmetic.

Teachers in kindergarten through grade six need a command of the mathematics for algebra and geometry described in this framework. Further, before entering a credential program, teaching candidates for kindergarten through grade six should have a full year of mathematics content courses that cover at least the material described in Chapter 3, "Grade-Level Considerations."

Junior high school and middle school mathematics teachers need a command of mathematics beyond that of kindergarten through grade six teachers. Before entering a credential program, teacher candidates for junior high school and middle school mathematics should have at least 24 semester hours of courses that are a part of an approved kindergarten through grade twelve mathematics credential program.

Before entering a credential program, high school mathematics teacher candidates need the full background required for state secondary certification in mathematics. College and university mathematics departments should design their programs so that their students majoring in mathematics do not encounter unnecessary obstacles in meeting the state requirements.

The Use of Technology

Technology has changed many aspects of doing mathematics, and mathematics education must reflect this reality. Further, some of technology's potential to enhance the teaching and learning of mathematics is already realized and appreciated.

However, along with the potential of such a powerful tool for doing good, the possibility also exists for doing immense, perhaps incalculable harm. There was a time during the past two decades when this critical message was not properly heeded. Some educators welcomed technology as the proverbial magic bullet. For example, the *Curriculum and Evaluation Standards for School Mathematics,* published by the National Council of Teachers of Mathematics (NCTM) (1989), states on page 8: "Contrary to the fears of many, the availability of calculators and computers has expanded students' capability of performing calculations. There is no evidence to suggest that the availability of calculators makes students dependent on them for simple calculations."

Likewise, it has been stated that technology in the classroom can be a positive force for equity because it helps break down barriers to mathematical understanding created by differences in computational proficiency. Some went so far as to recommend that in every grade calculators be issued to students just as textbooks are.

Such trust and optimism notwithstanding, more recent anecdotal evidence and large-scale studies offer a more sobering perspective. Over the years evidence has accumulated showing the ill-effects of calculator use in the schools. Among the more conservative observations is the following one, which was written by the mathematician-educator Leon Henkin (e-mail, March 20, 1996) in describing to his colleagues his experience as a volunteer in the mathematics classes of a local high school:

> When I first saw what [the graphing calculator] can accomplish, I was awed and excited. . . . However, after having spent three or four weeks in the class and seeing how, in practice, the calculator is actually used in the class, I have now concluded that it is about the largest obstacle to their gaining an understanding of the mathematical ideas of the course. The reason is that they have come to rely completely on the calculator to do arithmetic, as well as elementary algebraic calculations. If you ask them to estimate the slope of a function at a certain point when they are looking at the graph, they will punch in four numbers and calculate the difference quotient, and if you put your hand on the calculator to prevent them from doing so, they will assure you with all their might that they cannot multiply and divide without it.
>
> Obviously, this enormously stunts their ability to use graphs in an intuitive way to make conjectures and gain ideas about a problem whose solution they are seeking.

Beyond the anecdotal, there are also the findings from the Third International Mathematics and Science Study (TIMSS). For the eighth grade assessment, the majority (> 50%) of the students from three of the five nations with the top scores (Belgium, Korea, and Japan) never or rarely (once or twice a month) used calculators in mathematics classes.[1] In contrast, the majority of students (> 65%) from 10 of the 11 nations, including the United States, with scores below the international mean used calculators almost every day or several times a week in mathematics classes (Beaton et al. 1996). While this data does not prove that calculator use is damaging to the development of mathematical skills, it would be folly to ignore this correlation.

This kind of correlation and anecdotal evidence cannot be ignored because, as of 1998, almost no substantial body of reputable research or national studies exists to show how technology improves learning or whether in fact it does. The one large-scale study done to date on computer use is somewhat inconclusive (see "A Large-Scale Study of Computer Use" later in this chapter).

This framework proposes some guidelines on the use of technology in the classroom. The task is made difficult not only by an absence of hard data to help identify whether or when it is appropriate to use technology in mathematics education but also by the phenomenally rapid rate at which technology is advancing. For these reasons the scope of these guidelines will be restricted to what is warranted by the available evidence.

The Use of Calculators

The *Mathematics Content Standards for California Public Schools* was prepared with the belief that there is a body of mathematical knowledge—independent of technology—that every student in kindergarten through grade twelve ought to

---
[1] In the fourth country, Singapore, when the TIMSS was conducted, students used calculators rarely or not at all until the eighth grade.

> Almost no substantial body of reputable research or national studies exists to show how technology improves learning or whether in fact it does.

know and know well. Indeed, technology is not mentioned in the *Mathematics Content Standards* until grade six. More important, the STAR assessment program—carefully formulated to be in line with the standards—*does not allow the use of calculators all through kindergarten to grade eleven.*

Development of Basic Mathematical Skills

It is important for teachers to stay focused on the goal of having their students understand mathematics and develop the ability to use it effectively. Teachers should realize that understanding basic concepts requires that the students be fluent in the basic computational and procedural skills and that this kind of fluency requires the practice of these skills over an extended period of time (Bahrick and Hall 1991; Cooper and Sweller 1987; Sweller, Mawer, and Ward 1983). The extensive reliance on calculators runs counter to the goal of having students practice using these procedures. More to the point, it is imperative that students in the early grades be given every opportunity to develop a facility with basic arithmetic skills without reliance on calculators. For example, it should not be the case that, in a fifth-grade classroom, the simple addition $5/11 + 3/5$ can be done only with the help of a calculator. Students are expected to learn to be adept at making mental calculations and estimates. With a strong technical foundation in place, they will be in a good position to acquire the more complex concepts and procedures throughout their school years.

Preparation for the Use of Calculators

It should not be assumed that caution on the use of calculators is incompatible with the explicit endorsement of their use when there is a clear reason for such an endorsement. Once students are ready to use calculators to their advantage, calculators can provide a very useful tool not only for solving problems in various contexts but also for broadening students' mathematical horizons. One of the most striking examples of how calculators can be appropriately used to help solve problems is the seventh grade topic of compound interest. Initially, compound interest should be developed with problems that can be done without calculators, such as finding the total earnings over two or three time periods when the interest is compounded. However, once the general formulas are introduced, calculators become invaluable in answering questions, such as, What is the size of payments on a 20-year, fixed-rate loan of $50,000? or What will be the total amount of interest paid? (For more details see the discussion in Appendix C, "Middle School Sample Lesson: Compound Interest.")

Calculators can also be used in many situations to augment one's ability to teach important mathematical concepts. One can now include such problems as factoring 14,478,750 or finding the sum of $1/18731 + 11/2136$ in a regular sixth- or seventh-grade mathematics course with the expectation that students will recognize that such problems require the use of calculators. From the point of view of learning mathematics, such problems, particularly the first one, are not mere drill. They require a nontrivial use of systematic search procedures, which cannot easily be demonstrated with simpler problems. Another example is for students to seek the formula for the sum of squares of the first n positive integers. They might ask themselves if the desired formula could be a cubic polynomial in n. Calculators can also be invaluable in working with the extraction of difficult roots and in performing some calculations dealing with special functions, such as the exponential, logarithmic, and trigonometric ones.

Students must be facile in the execution of basic arithmetical and mathematical procedures without the use of calculators. As described in Chapter 1, "Guiding Principles and Key Components of an Effective Mathematics Program," basic computational and procedural skills influence employability in the United States (Rivera-Batiz 1992), and the practice of these skills contributes to the students' development of conceptual understanding and to the sophistication of students' problem-solving approaches (Siegler and Stern 1998; Sophian 1997).

The Use of Computers

The role of computers and software in learning and applying mathematics is changing almost as quickly as the technology is advancing. However, it is important to recognize that working at the computer cannot be considered a substitute for teachers' or for students' active individual involvement in a mathematics class or with homework assignments.

To help ensure that computer education is of maximal benefit to students, it is important to distinguish between general computer education and literacy and the applications of computers in the mathematics curriculum.

All students need to learn basic computer skills, and they should become familiar with several computer applications. They must have the ready ability to learn how to use computer applications and how to navigate and make use of the Internet. These skills can be taught starting in the elementary grades. It is extremely important, however, to recognize a good course in computer skills for what it is and not mistake it for a good course in mathematics. In particular, the teaching of computer literacy *should not take time away from, or be confused with, the teaching of the mathematics curriculum,* as described elsewhere in this framework.

Computers in the Elementary Grades

The large-scale use of computers *in the mathematics curriculum* needs to be viewed with extreme caution in kindergarten through grade five. (Some of the risks are detailed more fully in the report by H. Wenglinsky [1998], from the Educational Testing Service, discussed later in this chapter.) As noted previously, it is crucial for students in the elementary grades to acquire basic skills in arithmetic so that they can obtain a solid grounding in the concepts involved. These skills and concepts need to be solidly in place before students can take real advantage of computer or calculator applications in mathematics. Therefore, computers should not play a major role in the mathematics curriculum in kindergarten through grade five.

Computers in the Middle Grades

In the middle grades some schools may wish to take advantage of the many computer programs that can augment students' mathematical education. Here the schools and the individual teachers must be cautious, however. There are many excellent programs, rich in graphics and sound, that can help students learn basic concepts, such as fractions, and that can demonstrate graphs and even help with statistical and data analysis. But many programs are not appropriate for use in a mathematics curriculum. Some are more entertaining than educational, and others essentially do the mathematics for the students, thereby depriving them of the chance to learn for themselves.

For these reasons schools and teachers must judge carefully in deciding which types of computer programs to use to augment their mathematics curriculum. Those computer programs must be chosen with the same care as that used in choosing other types of instructional materials, such as textbooks. The decision-making process should always focus on whether the program augments the students' learning of the curriculum.

Computers in High School

In the higher grades, students' levels of interaction with computers can be more sophisticated. In particular, computer-programming skills are based on the same logical foundations as is mathematics itself and, if possible, courses in mathematical computer programming should be available to students. However, it is not appropriate to discuss the details of such a course here.

Within the mathematics curriculum the writing of simple computer programs to solve problems in mathematics (or science) can be encouraged through extra-credit homework assignments or projects. The advantages of such assignments are as follows:

- The writing of a program to perform a mathematical calculation can help to solidify students' knowledge of the mathematical concepts involved in solving the problem.
- The lack of tolerance for errors in computer programs forces the students to be precise and careful in their work.
- The algorithms that students implement must be correct and correctly expressed in the computer language being used.

All these benefits are important, not only for the students' mathematical development but also in many other areas. As always, caution needs to be emphasized.

For example, merely entering data into a prewritten program does not have much educational benefit and can easily be counterproductive.

Computer skills, particularly the programming skills described previously, can give students a real advantage in entering the job market. Thus, serious attention must be given to developing appropriate courses. Merely being familiar with how to use word-processing programs or interactive mathematics programs will not achieve these objectives.

A Large-Scale Study of Computer Use

Recently, the first large-scale study of the relationship between computer use and student achievement was done by H. Wenglinsky (1998), under the auspices of the Educational Testing Service. He used data from the 1996 National Assessment of Educational Progress (NAEP) examinations and the questionnaires filled out by the fourth and eighth grade students taking the examinations, their teachers, and the administrators of the schools involved. The basic conclusions, appearing on page 30 of the report, give *correlations* between the various ways in which computers were used and the corresponding student scores on the NAEP examinations.

These correlations are expressed in weeks out of an average school year. Thus, a correlation of -0.45 would mean that, on average, a student who used computers in a particular way would score at the same level on the NAEP examination as would an average student who had had 16.2 weeks *less schooling* in a 36-week school year, as shown in the following equation:

$$16.2 = 0.45 \times 36$$

The largest correlation in the study, -0.59 was for eighth graders who used computers for drill. They *lagged behind the average student* on the NAEP examinations. The −0.59 denotes that they trailed the average by 21.2 weeks out of an average 36-week school year. In contrast, eighth graders who used computers for applications of mathematics showed a gain of 0.42, or 15.1 weeks, in their average scores.

Fourth graders who used computers for mathematics-related games in school showed a positive correlation of 0.15, or 5.4 weeks *ahead* of the average. But there was an even larger negative correlation between overall school computer use by fourth graders and lower student achievement (−0.20, or 7.2 weeks behind the average). The biggest correlation of all for fourth graders was −0.26, or 9.4 weeks behind the average, for students who used computers at home.

The remaining measures were significantly smaller, were related to variables such as socioeconomic status (over which the schools have no control), or were involved with teacher preparation.

Note: As with the TIMMS data presented earlier in this chapter, these interpretations of the NAEP data do not necessarily relate to cause and effect. It could well be that using the computer for drill is a symptom, not a cause, of student difficulties. But, in general, the results of this study strongly reinforce the concerns about overdependence on the computer as a teaching aid.

The Use of the Internet

The Internet is another technological advance that offers interesting opportunities for student learning in mathematics. It provides a wealth of information more rapidly than was possible at any other time in human history. One way teachers might take advantage of this technology is to require students to obtain data for statistical analysis from the Net. But at this stage the Internet seems more useful to teachers themselves than to students. A great deal of information on lesson plans and curricula is available through the Internet, as are many opportunities for teachers to exchange information and advice.

Mathematics teachers should exercise caution to ensure that using the Internet promotes student learning and is an efficient use of student time. If the Internet is a feature of a mathematics lesson, its use must be directly related to the goal of the lesson, and the information accessed must be directly related to the mathematical content of the lesson.

The use of technology in and of itself does not ensure improvements in student achievement, nor is its use necessarily better for student achievement than are more traditional methods. Only carefully conducted research studies will determine how, and with what mathematical content, technology will foster student achievement. At this time the research base is insufficient for making any strong recommendations in favor of using computer and calculator technology in kindergarten through grade six classrooms. Consequently, the framework recommends that extreme care be used in adopting any instructional programs based solely on the use of computers or calculators.[2]

---

[2] This statement does not refer to programs that use electronic technology as an alternative means to deliver instructional material.

10

Criteria for Evaluating Instructional Resources

Instructional materials adopted by the state must help teachers present the content set forth in the *Mathematics Content Standards*. To accomplish this purpose, this chapter establishes the criteria for evaluating the instructional materials. These criteria serve as evaluation guidelines for the statewide adoption of mathematics instructional materials in kindergarten through grade eight; they may also provide guidance for the development and review of instructional materials for grades nine through twelve.

The California mathematics standards are challenging. In the initial years of implementing the framework, a major goal for most school districts across the state will be to facilitate the transition from what students actually know to what the *Mathematics Content Standards* envisions they should know. Instructional materials play a central role in facilitating this transition.

Publishers are encouraged to design instructional materials specifically for use during this transition period. Materials that will help districts meet this challenge will clearly identify the highest-priority instructional activities and will allow teachers to focus instruction in those areas as necessary. During this transition school districts may need to allocate more time to mathematics instruction than they will in subsequent years.

The California State Board of Education adopted instructional materials that, on the whole, should provide programs that will be effective for all students—those who have not mastered most of the content taught in the earlier grades and those who have. In addition, some instructional materials must specifically address the needs of teachers who instruct a diverse student population. Therefore, the framework does not ask publishers to use a particular pedagogical approach; instead, it encourages them to select research-based pedagogical approaches that collectively give teachers alternatives that will help them in teaching mathematics effectively.

The criteria are organized into five general categories, followed by a section on suggestions for optional criteria for publishers who choose to develop transition materials:

1. *Mathematical content/alignment with the standards.* The content as specified in the *Mathematics Content Standards* and elaborated on in this framework
2. *Program organization.* The sequence and organization of the mathematics program that provides structure to what students should learn each year
3. *Assessment.* The strategies presented in the instructional materials for measuring what students know and are able to do
4. *Universal access.* The information and ideas that address the needs of special student populations, including students eligible for special education, students whose English-language proficiency is significantly lower than that typical of their class or grade level, students whose achievement is either significantly below or above that typical of their class or grade level, and students who are at risk of failing mathematics
5. *Instructional planning and support.* The instructional planning and support information and materials, typically including a separate edition specially designed for use by the teacher, that help teachers in implementing the mathematics program

Mathematics Content/Alignment with the Standards

Mathematics materials should support teaching to the mathematics content standards. In kindergarten through grade seven, the standards are organized in five strands: Number Sense; Algebra and Functions; Measurement and Geometry; Statistics, Data Analysis, and Probability; and Mathematical Reasoning. However, there is no requirement that publishers adhere to this strand organization as long as they address all the individual standards. In grades eight through twelve, the standards are organized by discipline. Some schools teach the grade eight through twelve mathematics curriculum in traditional courses, and others teach it in an integrated fashion. To provide local educational agencies and teachers with flexibility in presenting the material, the standards for grades eight through twelve do not mandate that a particular discipline be initiated and completed in a single grade. Nevertheless, however mathematics is taught, the core content of these subjects must be covered; and all academic standards for achievement must be the same.

Materials that fail to provide thorough instruction on the standards and the mathematics content described in this framework will not be considered for adoption.

Materials aligned with the mathematics content standards must satisfy the following criteria:

- The content supports teaching the mathematics standards at each grade level (as detailed, discussed, and prioritized in Chapters 2 and 3 of this framework).
- A checklist of evidence accompanies the submission and includes page numbers or other references and demonstrates alignment with the mathematics content standards and, to the extent possible, this framework.
- Mathematical terms are defined and used appropriately, precisely, and accurately.
- Concepts and procedures are explained and are accompanied by examples to reinforce the lessons.
- Opportunities for both mental and written calculations are provided.
- Many types of problems are provided: those that help develop a concept, those that provide practice in learning a skill, those that apply previously learned concepts and skills to new situations, those that are mathematically interesting and challenging, and those that require proofs.
- Ample practice is provided with both routine calculations and more involved multistep procedures in order to foster the automatic use of these procedures and to foster the development of mathematical understanding, which is described in Chapters 1 and 4.
- Applications of mathematics are given when appropriate, both within mathematics and to problems arising from daily life. Applications must not dictate the scope and sequence of the mathematics program, and the use of brand names and logos should be avoided. When the mathematics is understood, one can teach students how to apply it.
- Selected solved examples and strategies for solving various classes of problems are provided.
- Materials must be written for individual study as well as for classroom instruction and for practice outside the classroom.
- Mathematical discussions are brought to closure. Discussion of a mathematical concept, once initiated, should be completed.
- All formulas and theorems appropriate for the grade level should be proved, and reasons should be given when an important proof is not proved.
- Topics cover broad levels of difficulty. Materials must address mathematical content from the standards well beyond a minimal level of competence.
- Attention and emphasis differ across the standards in accordance with (1) the emphasis given to standards in Chapter 3; and (2) the inherent complexity and difficulty of a given standard.
- Optional activities, advanced problems, discretionary activities, enrichment activities, and supplemental activities or examples are clearly identified and are easily accessible to teachers and students alike.
- A substantial majority of the material relates directly to the mathematics standards for each grade level, although standards from earlier grades may be reinforced. The foundation for the mastery of later standards should be built at each grade level.
- An overwhelming majority of the submission is devoted directly to mathematics. Extraneous topics that are not tied to meeting or exceeding the standards, or to the goals of the framework, are kept to a minimum; and extraneous material is not in conflict with the standards. Any nonmathematical content must be clearly relevant

to mathematics. Mathematical content can include applications, worked problems, problem sets, and line drawings that represent and clarify the process of abstraction.
- Factually accurate material is provided.
- Principles of instruction are reflective of current and confirmed research.
- Materials drawn from other subject-matter areas are scholarly and accurate in relation to that other subject-matter area. For example, if a mathematics program includes an example related to science, the scientific references must be scholarly and accurate.
- Regular opportunities are provided for students to demonstrate mathematical reasoning. Such demonstrations may take a variety of forms, but they should always focus on logical reasoning, such as showing steps in calculations or giving oral and written explanations of how to solve a particular problem.
- Homework assignments are provided beyond grade three (they are optional prior to grade three).

Program Organization

The sequence and organization of the mathematics program provide structure to what students should learn each year and allow teachers to convey the mathematics content efficiently and effectively. The program content is organized and presented in a manner consistent with achieving the goals of the mathematics content standards. The essential components for program organization are listed as follows:

- Concepts are developed in logical order and increase in depth and complexity during each school year and from grade to grade. Materials for each grade are organized around a few key topics, as described in Chapter 3 of this framework. Although some repetition in the form of review is necessary, review must be for developing automaticity or preparing for further learning. Content for a grade level must not be diluted by an extensive review of skills that have been covered earlier. Substantial new material needs to be introduced at successive levels.
- The order of presentation of mathematical topics is mathematically and pedagogically sound.
- Prerequisite skills and ideas are presented before the more complex topics that depend on them.
- Coverage starts with easy cases and proceeds, step-by-step, to increasingly complex problems within the topic areas.
- The connections between related topics are taught when it is appropriate, and the organization of the material supports the understanding of these connections.
- Mathematical content and instructional activities are sequenced to prevent common student misconceptions (see Chapter 3). Topics that students are likely to confuse are not introduced together, but similarities and differences in ideas and procedures are eventually addressed.
- Student materials ensure that students can look back in the textbook for help with understanding a topic; compilations, such as indices, tables of contents, and review summaries, also provide assistance.
- Materials include tables of contents, indices, and glossaries containing important mathematical terms used in the book to make it easier for parents or others to tutor students. The framework encourages any features of instructional materials that enable older sibling, parental, or other adult tutoring.

- The scope and sequence are referenced in such a way that "looking back and forward" can include previous and subsequent grade levels in the series.
- Materials include an overview of chapters that students are expected to learn with the mathematical concepts involved clearly identified. This material should be available to students, parents, and teachers.
- Problems and exercises based on the students' prior and current experience with the mathematics curriculum are accessible to students.
- Materials are designed so that if students should have trouble with a particular type of problem, guidance is provided to the teachers to help them identify the reason for the difficulty (e.g., identify component skills not mastered), and specific remedies should be suggested.
- Support materials, such as computer programs and manipulatives, are clearly aligned with the mathematical and instructional goals of the mathematics content standards and this framework.
- Applications of the mathematics under discussion must be clearly marked as such and must not be equated with the mathematics itself.
- Materials introduce new concepts at a reasonable pace and provide sufficient instructional and practice material on all the important topics.
- Standards-based goals are explicitly and clearly associated with instruction and assessment.
- Computational and procedural skills, conceptual understanding, and problem solving are interconnected and included throughout the program.

Assessment

Instructional materials should contain strategies and tools for continually measuring student achievement with a reasonable degree of accuracy. Assessments will measure what students know and how well they know it.

This framework addresses assessment in Chapter 5. In keeping with the issues discussed in that chapter, instructional materials must provide teachers with a variety of assessment measures and procedures for different purposes. Assessment programs should include elements of conceptual understanding, basic and procedural skills, and problem solving.

Instructional materials should include:

- Assessments that have content validity and measure individual student progress at regular intervals, that measure each student's entry-level skills and knowledge, that monitor student progress toward meeting the standards, that evaluate mastery of grade-level standards, and that provide summative evaluations of individual student achievement
- Assessments for identifying students who are not making reasonable progress toward achieving the standards
- Opportunities to assess student reasoning across the grades as it progresses from informal explanation to formal proofs
- Measurement of conceptual understanding, basic skills and procedures, and problem solving

Instructional materials should provide a variety of assessment measures and procedures for different purposes, including:

- Assessments that are appropriate for different grade levels so that students can check their own work frequently while learning the material and after completing a chapter or unit
- Assessment of appropriate duration at various intervals (e.g., every day, at the end of a lesson or chapter, and at intervals of no more than six weeks)

- Research-based assessments that have content validity
- Both curriculum-embedded assessment and summative assessment
- Multiple methods of assessing what students know and are able to do

Instructional materials must guide the teacher in assessing the student's level at the beginning of the school year. The initial assessment should be comprehensive and help the teacher in determining whether the student should work with the grade-level materials, the materials for the previous grade level, or the transitional materials that teach concepts and skills that should have been previously mastered.

Instructional materials should help teachers use assessment data in instructional planning and reporting, such as:

- Suggestions based on assessment data about ways in which to modify an instructional program so that all students are constantly progressing toward meeting or exceeding the standards
- Suggestions about the type of assessment data to be used to guide decisions about instructional practices
- Suggestions for keeping parents and students informed about student progress

Universal Access

Instructional materials need to provide access to the standards-based curriculum for students with special needs. Programs must conform to the policies of the California State Board of Education and to other applicable state and federal requirements with respect to diverse populations and students with special needs, as discussed in Chapter 6, "Universal Access."

Materials supporting universal access include:

- Strategies to help the teacher provide access to mathematics for all students with regard to ability, language proficiency, and other special needs[1]
- A description of methods by which special needs students can experience success with and appreciation of mathematics, from the simplest skills to the most complex understanding
- Help for teachers to offer the program to students with a wide range of achievement levels, making suggestions for compacting or expanding the curriculum and grouping within or across grade levels
- Help for students who are below grade level, including more explicit explanations, review, practice, guidance, or other assistance (These students will need extra time and instructional materials devoted to mathematics. It is also important that accommodations for special needs or other low-performing students provide opportunities for them to learn the key concepts in mathematics and not relegate struggling students to meaningless tasks.)
- Alternatives for gifted and talented students that are thoughtful and well conceived and that allow students to accelerate beyond their grade-level content (acceleration) or to study the content in the *Mathematics Content Standards* in greater depth or complexity (enrichment)
- Information about how teachers might use the results of assessment to differentiate curriculum and instruction at the appropriate level of challenge for all students

---

[1] *Note:* The California *Education Code* provides for adopted instructional materials to be translated into braille and large print by the Clearinghouse for Specialized Media and Technology. The Clearinghouse also converts materials into tape and video format as appropriate.

- Suggestions to help teachers preteach and reinforce mathematics vocabulary and concepts with English learners
- Suggestions to teachers on how and when to modify assessment or instruction for special education pupils

Instructional Planning and Support

Materials that provide support for teachers need to be built into the program. These materials should contain specific suggestions and illustrative examples of how the teacher can implement a standards-based mathematics program. Instructional materials should meet the following criteria:

- All components of the program are provided so that there is little or no need for teachers to identify, gather, or develop supplementary materials.
- Clear grade-appropriate explanations of mathematical concepts appear in a form that teachers can easily adapt for classroom presentation.
- (Optional) Teacher resources contain full, adult-level explanations and examples of the more advanced mathematics topics that relate to the lesson so that teachers can assess and improve their own knowledge of the subject as necessary. (East Asian lesson plans offer excellent examples showing how this can be done; see Appendix B.)
- Teacher resources contain discussions of the role of the specific grade-level mathematics in relation to the total kindergarten through grade twelve mathematics curriculum and beyond, describing both what has been previously taught and why and what will be taught in succeeding grades.
- Different kinds of lessons and alternative ways in which to explain concepts are provided to offer teachers choice and flexibility in developing their programs.
- Any required manipulatives are provided, or inexpensive alternatives are suggested.
- Manipulatives should promote student learning, and clear instructions for their efficient use are provided.
- Teacher materials contain sample lesson plans and suggestions for organizing and managing the classroom.
- Tools for assessing student progress and knowledge and suggestions for how to use the assessment data for instructional planning are provided.
- A system is provided for accelerating or decelerating the rate at which new material is introduced to students, in accordance with students' ability to assimilate new material.
- Review and practice distributed over time, as described in Chapter 4, is provided to enhance understanding and promote generalization and transfer of skill and knowledge.
- Any instructional software and technological tools used as a format for presentation of the instructional materials are an integral part of the submission.

Special Consideration: Support for Teachers During the Transition Period

The California mathematics standards aim at a level considerably above that which many students had achieved when this framework was written. Helping students make the transition to the levels of the standards requires a major effort. During the first two or three years of transition, or perhaps longer, a sixth grade teacher, for example, will most likely use an instructional program aimed at helping many students whose performance level falls far short of the grade-level standards to catch up. In subsequent years, that teacher may need to use *the same instruc-*

tional program to maintain and expand on the grade-level performance for students who enter the sixth grade already performing at grade level.

Instructional materials should provide a program that will be effective for all students—those who have not mastered most of the content taught in the earlier grades and those who have. Some students may have weaknesses in several areas of content from the earlier grades. This material can be taught within the context of the grade-level textbook. Other students may have such severe problems that it would be unrealistic to assume that the deficits could be remediated with the grade-level textbook.

The hope is that some publishers will directly address the need for transitional materials designed to help students reach the levels of proficiency required in the *Mathematics Content Standards*. Such transitional materials may be designed for a two-hour block of mathematics instructional time per day, a summer or "off-track" program, or an after-school tutoring program of up to one hour per day. Those publishers should provide transitional materials with content related to the standards, techniques for assessment, and support for teachers. Those topics are discussed in the next sections.

Content Related to the Standards

A standards map should be provided showing which standards are addressed and when, with the understanding that the transition materials include standards from several grade levels in a single student or teacher edition. Publishers may consider including transition materials designed to teach the essential content from earlier grades along with the standards for a given grade.

Assessment Materials

Assessment materials should be provided to help the teacher determine the student's level of achievement relative to the standards at the beginning of the school year. The initial assessment should be comprehensive so that the teacher can determine which textbook would be appropriate for the student:

- The grade-level textbook
- The grade-level textbook for a previous grade
- Special transitional materials that teach concepts and skills that should have been mastered earlier

Teacher Support

Suggestions for teaching students lacking knowledge of certain content cannot be simple afterthoughts to the grade-level material. To develop appropriate instructional plans for these students, teachers need a master guide that enables them to identify foundational skills and associated instructional units taught at earlier grade levels. Materials for students functioning below their grade levels must be designed to accelerate the students' acquisition of critical concepts, procedures, and skills. Another consideration in the development of these materials is that more than one hour a day of instructional time may be devoted to mathematics for students in grades four through twelve who are not performing at grade level. Instructional programs should provide teachers with instructional activities for use during any additionally allocated instructional time. Placement tests and suggestions for instructional strategies should be included to help students whose facility with mathematics enables them to move through the program at an accelerated pace.

Appendix A
Sample Instructional Profile

Instructional Profile: Adding and Subtracting Fractions with Unlike Denominators, Grades Four Through Seven

Adding and subtracting fractions with unlike denominators has been selected for elaboration because:

- The student is required to use a number of discrete skills to solve these problems.
- There are specific types of problems within this category, each with its own particular prerequisite, and some problems are much more complex than others.

These types of problems and the prerequisites for solving them need to be introduced and practiced for varying amounts of time. The teacher has to be adequately prepared to present each concept for solving problems when the students are ready for it.

Standards for Adding and Subtracting Fractions

The standards in this section present the content and the related concepts and skills that students need to be able to add and subtract fractions with unlike denominators.

Grade 5. Number Sense 2.3. Solve simple problems, including ones arising in concrete situations, involving the addition and subtraction of fractions and mixed numbers (like and unlike denominators of 20 or less), and express answers in the simplest form.

Grade 6. Number Sense 2.1. Solve problems involving addition, subtraction, multiplication, and division of positive fractions and explain why a particular operation was used for a given situation.

Related standards. To solve the problems required by the content standards listed previously, students need to learn related concepts and skills:

- Compute and perform simple multiplication of fractions.
- Rewrite a fraction as an equivalent fraction with a different denominator.
- Reduce fractions.
- Convert improper fractions to mixed numbers.
- Determine the operation called for in a story problem or in a concrete situation in which fractions appear and create a problem to be solved.

Considerations for Instructional Design

The following considerations for instructional design are discussed in this section: sequence of instruction, teaching the components of complex applications, selection of examples, introduction of concepts and skills, scaffolded instruction, and practice.

Sequence of Instruction

This section presents procedures and cautions for teachers to follow when they introduce addition and subtraction of fractions with unlike denominators.

1. Students often confuse similar concepts. For example, finding the greatest common factor and finding the lowest common multiple are potentially confusing because they both deal with multiples of numbers. This potential confusion can be decreased by not introducing the concepts in close proximity.

2. Easier problems are to be taught before more difficult ones. While this guideline seems self-evident, its application requires identifying subtypes of a general set of problems. For example, problems that require borrowing ($2\frac{1}{2}-1\frac{3}{4}$) are slightly more difficult than problems that do not require borrowing ($2\frac{3}{4}-1\frac{1}{2}$).

   There are a number of distinct types of problems for adding and subtracting fractions with unlike denominators:

   - Simple problems in which one fraction has the common denominator: $\frac{1}{2}+\frac{3}{4}$
   - Simple problems in which the common denominator is neither denominator: $\frac{3}{4}-\frac{1}{5}$
   - Problems with mixed numbers—no regrouping is required: $5\frac{1}{5}+3\frac{2}{3}$
   - Problems that require adding three fractions: $\frac{1}{3}+\frac{1}{5}+\frac{2}{3}$
   - Problems that require subtracting a mixed number from a whole number ($8-3\frac{1}{4}$) and adding mixed numbers: ($7\frac{3}{4}+2\frac{4}{5}$)
   - Mixed-number problems that require regrouping ($5\frac{1}{2}-2\frac{3}{4}$) and rewriting the numerator and denominator.

3. Introducing too much information in a single lesson or over a short period of time can result in student confusion. The introduction of new skills and applications needs to be controlled so that children at a particular skill level can reasonably absorb the new information.

Teaching the Components of Complex Applications

Component skills need to be thoroughly taught before they appear in complex applications. For example, the concept that a fraction equals one when it has the same numerator and denominator is a very important component for finding equivalent fractions and for converting mixed and improper fractions.

Example: Fractions That Equal One

This concept is included in the standards for the earlier grades. The concept of a fraction equaling one is applied in:

- *Equivalent fractions.* Illustrative explanatory wording follows: "When you multiply by one, you get the same quantity you started with, so when you multiply by a fraction that equals one, you get a quantity equal to what you started with."

$$\frac{3}{4} \times \boxed{\frac{2}{2}} = \frac{6}{8}$$

- *Converting improper fractions to mixed numbers.* Initial teaching can show students how to rewrite an improper fraction as an addition problem of fractions equal to one and a fraction less than one.

$$\frac{13}{5} = \boxed{\frac{5}{5}} + \boxed{\frac{5}{5}} + \frac{3}{5} = 2\frac{3}{5}$$

- *Reducing fractions.* A sophisticated strategy for reducing involves rewriting the numerator and denominator as prime factors, then crossing out fractions that equal one:

$$\frac{12}{18} = \frac{2 \times 2 \times 3}{2 \times 3 \times 3} = \frac{2}{3}$$

Selection of Examples

Examples in teaching sets should be designed and selected to rule out possible misinterpretations. For example, if a teacher teaching students to solve application problems containing the words "younger than" presents only subtraction problems as examples, the students might develop the idea that the phrase "younger than" always involves subtraction. (*Jan is 19 years old. Alice is 7 years younger than Jan. How old is Alice?*) The students would likely miss a problem that calls for addition, such as *Marcus is 25 years old. He is 7 years younger than his brother. How old is Marcus's brother?*

Teachers should not assume that all students will automatically be able to generalize to new types of problems. For example, when students start working on new types of verbal problems requiring fractions, the teacher should point out the following facts: (1) the problems differ from problems they are familiar with; (2) the problems involve fractions; and (3) they must be careful when setting up the problems.

Example: Concrete Applications

Number Sense Standard 2.3 for grade five states that students are to "solve simple problems, including ones arising in concrete situations, involving the addition and subtraction of fractions and mixed numbers (like and unlike denominators of 20 or less), and express answers in the simplest form."

Once children have learned to compute answers to problems involving the addition and subtraction of fractions and have had sufficient practice to work the problems without prompting, applications such as the following can be introduced:

- *The recipe calls for $\frac{3}{8}$ of a pound of nuts. Anna has only $\frac{1}{4}$ of a pound of nuts in her kitchen. If she goes to the store to buy nuts, what fraction of a pound of nuts will she need?*
- *There is $\frac{2}{3}$ of a pizza left over, and $\frac{1}{4}$ of another equal-sized pizza is left over. If we put the pieces from both pizzas together, what part of a whole pizza would we have?*

Introduction of Concepts and Skills

Initial teaching should be interactive. The teacher not only demonstrates and explains but also asks frequent questions to check for understanding.

Example: Initial Strategy for Adding and Subtracting Fractions with Unlike Denominators

It must be emphasized that the basic definition of the adding or subtracting of fractions is simple and direct. It is only the refinements of the definition that cause complications. Students should be told that the *definition is more basic than the refinements.*

The basic *definition* is to multiply the denominators of the fractions to make a common denominator. For example, the addition problem of $\frac{3}{4}+\frac{2}{5}$ is done in the following way to obtain the answer $\frac{23}{20}$:

$$\frac{3}{4} + \frac{2}{5}$$

$$\frac{3}{4} \times \left[\frac{5}{5}\right] + \frac{2}{5} \times \left[\frac{4}{4}\right]$$

$$= \frac{15}{20} + \frac{8}{20}$$

$$= \frac{23}{20}$$

Sometimes, it is possible by inspection to decide on a common denominator of two fractions that is smaller than the product of the two denominators. Thus $\frac{5}{6}+\frac{1}{8}$ equals $\frac{23}{24}$, because visibly 24 is a common multiple of the two given denominators 6 and 8. One proceeds to rewrite the two fractions with 24 as the denominator:

$$\frac{5}{6} \times \left[\frac{4}{4}\right] + \frac{1}{8} \times \left[\frac{3}{3}\right]$$

$$\frac{20}{24} + \frac{3}{24} = \frac{23}{24}$$

However, students should also be told that if they add $\frac{5}{6}+\frac{1}{8}$ using the basic definition (see p. 240) to obtain $\frac{5}{6}+\frac{1}{8}=\frac{40}{48}+\frac{6}{48}=\frac{46}{48}$, then the resulting answer $\frac{46}{48}$ *is as valid as* $\frac{23}{24}$. (Note that $\frac{46}{48}$ would be incorrect only if a reduced fraction is specifically requested for the answer.)

Again, if the basic definition of the subtraction of fractions is used, we would obtain an *equally valid answer*:

$$\frac{13}{42} - \frac{2}{35} = \frac{13 \times 35}{42 \times 35} - \frac{2 \times 42}{35 \times 42}$$

$$= \frac{455}{1470} - \frac{84}{1470} = \frac{371}{1470}$$

$$\frac{13}{42} - \frac{2}{35} = \frac{13}{2 \times 3 \times 7} - \frac{2}{5 \times 7}$$

$$= \frac{13}{2 \times 3 \times 7} \times \boxed{\frac{5}{5}} - \frac{2}{5 \times 7} \times \boxed{\frac{2 \times 3}{2 \times 3}}$$

$$= \frac{65}{210} - \frac{12}{210} = \frac{53}{210}$$

The advantage of the least common multiple procedure therefore lies in avoiding the computations with relatively large numbers.

In context, any one of these strategies may be selected for use if it happens to be the most convenient. Students should always be taught the basic definition (see p. 240) first, but the order of teaching the other strategies can vary. Ultimately, they must know all three.

Scaffolded Instruction

Teachers should not expect students to make the jump from observing the teacher work a problem to being able to work problems themselves independently. Teachers need to provide support with teacher/student questions and feedback and for prompted written problems.

First Example of Scaffolded Instruction: Equivalent Fractions

In earlier grades students worked with diagrams to identify equivalent fractions. In the fifth grade a symbolic procedure is presented.

What follows is an illustration of teacher wording for scaffolded instruction:

$$\frac{2}{4} \times \frac{\boxed{}}{\boxed{}} = \frac{\square}{12}$$

The teacher tells the class to read the problem. Then the teacher says:

> We know the fractions are equal. We have to find a missing number in one of the fractions.
>
> We know that multiplying by one gives an answer equal to what we started with.
>
> Let's make a fraction equal to one.
>
> What number do we multiply by four to get 12?
>
> The bottom number of the fraction that equals one is three. So what must the top number of the fraction be?
>
> Yes, three-thirds equals one.
>
> Write the fraction of one, then multiply across the top and find the missing number.
>
> Two-fourths equals how many twelfths?
>
> Yes, two-fourths equals six-twelfths.

Second Example of Scaffolded Instruction: Adding and Subtracting Fractions with Unlike Denominators

The wording shown on the next page could be used after the teacher, using a highly prompted procedure for several days, has taught the students how to find the least common denominator. The teacher is now ready to decrease the structure while still providing guidance for the children.

The use of the diagram to prompt converting fractions makes the task easier initially. The extensiveness of the diagram prompts would gradually be decreased until students were able to do the problems without a diagram. After students can work problems written vertically, the teacher introduces problems written in a horizontal format. The teacher then prompts students on applying the same steps.

$$\frac{3}{4} \times = \frac{}{}$$
$$-$$
$$\frac{1}{8} \times = \frac{}{}$$

The teacher tells the class to read the problem, then says:

> Can we work the problem the way it is written?
>
> What do we have to do?
>
> Yes, rewrite the fractions as equivalent fractions with the same denominators.
>
> Rewrite the fractions and then add the new fractions.

Practice

A key element in preparing students to work independently is integrated practice in which newly introduced material appears along with previously introduced material that may be similar and is likely to be confused. For example, after teaching students to work subtraction problems, such as $4\frac{2}{3} - \frac{1}{4}$, the teacher includes such problems as $4\frac{2}{3} - \frac{5}{6}$ in which borrowing is required.

Assessment

Three kinds of assessment are discussed in this section: entry-level assessment, monitoring student progress toward instructional objectives, and post-test assessment toward the standard.

Entry-Level Assessment

Before students begin instruction on new content, they need to be proficient in the prerequisite skills and concepts taught in earlier grades. These prerequisites should have been reviewed frequently enough to facilitate retention. Still, assessment is required before instruction that is dependent on them begins.

- The prerequisites from the grade three content standards for adding and subtracting fractions with unlike denominators are:

Number Sense 3.1. Compare fractions represented by drawings or concrete materials to show equivalency and to add and subtract simple fractions in context (e.g., $\frac{1}{2}$ of a pizza is the same amount as $\frac{2}{4}$ of another pizza that is the same size; show that $\frac{3}{8}$ is larger than $\frac{1}{4}$). **Note: Students have worked with drawings and concrete materials to show equivalencies. They should not be tested on finding equivalencies without these prompts.**

Number Sense 3.2. Add and subtract simple fractions (e.g., determine that $\frac{1}{8} + \frac{3}{8}$ is the same as $\frac{1}{2}$).

- The prerequisite for working with adding and subtracting fractions with unlike denominators from the grade four standards are:

Number Sense 1.7. Write the fraction represented by a drawing of parts of a figure; represent a given fraction using drawings; and relate a fraction to a simple decimal on a number line.

Monitoring Student Progress Toward Instructional Objectives

As students are able to do each skill independently, the instructional program should include testing to determine student proficiency in what has been taught as skills are developed. The testing should be in the same form as the applications that students have been taught. Guidance should be provided to enable the teacher to determine the specific cause if a student is unable to work a problem correctly, and immediate remediation should be provided on that particular skill or procedure.

Post-test Assessment Toward the Standard

Post-tests include end-of-quarter, end-of-semester, standardized norm-referenced, and standards-based tests.

Considerations for Universal Access

Students with disabilities or learning difficulties need to be very carefully assessed to ensure that they have mastered all the earlier-taught skills needed to do the applications presented. For example, students doing equivalent fractions need to be able to figure out a problem such as $4 \times \_ = 20$.

Some skills are important throughout a grade level. For example, in the fifth grade knowledge of the multiplication facts is a critical component for many applications. Students who are slow at or who are unable to figure out multiplication facts would be at a great disadvantage. The assessment of multiplication facts should occur early in the school year so that there is ample time to teach the necessary facts.

Students with disabilities or learning difficulties often will require extra scaffolded instruction to apply more involved procedures and may require scaffolding as the new skills are integrated with previously taught skills. Also, students will benefit from immediate feedback as they work problems.

Advanced learners can be provided with substitute challenge problems to be worked individually or in a cooperative setting, or they can write about connections among concepts.

Teachers presenting lessons to English learners should use:

- As few terms as possible and still get the point across
- The most basic and accurate terms
- Those few, basic terms consistently

Teachers should give English learners extra opportunities to learn and master difficult but critical vocabulary. Teachers should not compromise on the mathematical content but should try to be careful in the use of English language applications.

Appendix B
Elementary School Sample Lesson: An East Asian Approach

Lesson One, "Introduction to Adding—The Adding-on Concept," is one of five lessons from Unit Five, "Addition with Sums Up to Ten," from "M-Math," an elementary school mathematics curriculum being developed by the University of Michigan and based on an East Asian curricular approach. This is just one lesson in a series covering five strands of mathematics concepts: number sense, operation and computation, measurement, geometry, and quantitative relations. These strands are introduced sequentially through in-depth discussion of each of the systemic laid-out mathematics concepts. The M-Math project is directed by Dr. Shin-ying Lee at the Center for Human Growth and Development, University of Michigan, with the help of many colleagues and teachers. It is an effort, based on what has been learned from the East Asian approach, to adapt and develop mathematics lessons to be taught in American classrooms. The project is supported by the Pew Charitable Trusts and the University of Michigan.

Unit Five
Addition with Sums Up to Ten

Lesson 1   Introduction to Addition—The Adding-on Concept
Lesson 2   The Combining Concept of Addition
Lesson 3   Both Types of Addition Problems
Lesson 4   Addition with Zero
Lesson 5   Vertical Addition

Introduction to the Unit

Up to this point, students should have learned and mastered the concept of numbers through ten. In this unit students will be introduced to two concepts of addition. The first is the *adding-on* concept, which involves gradually adding objects to an original group and finding the resulting quantity. The second type is the *combining* concept, or adding two separate groups together and finding the combined quantity. These concepts illustrate two related yet distinct situations in which addition is used.

Students will also learn how to write addition equations and to recognize and use key mathematical terms related to addition (e.g., *add, more, total, altogether*) in this unit. To facilitate students' understanding of the relationship between the addition concepts and the written equation, all addition problems are presented in the form of addition "stories." Moreover, students are expected to come up with their own story problems for each type of addition concept. At the end of this unit, students will learn how zero can be incorporated into addition equations as well.

Goals of the Unit

At the end of this unit, students should be able to:

- Understand the meaning of the two approaches to addition introduced in this unit:
  1. The adding-on concept
  2. The combining concept
- Apply addition to real-life examples in which the sum is ten or less.
- Learn and use the terms and symbols related to addition, such as *add, addition, equation,* +, and =.
- Read and write addition equations.
- Understand the meaning of addition equations that involve zero.

Essential Mathematical Concepts in Unit Five

The mathematics concepts presented in this unit are the meaning and applications of addition, important points to keep in mind when teaching addition, important aspects about equations, and addition involving zero.

The Meaning and Applications of Addition

Although addition is a straightforward mathematical operation, first grade students often demonstrate difficulties in its application. For example, students sometimes use addition in situations in which they should be using subtraction or vice versa. To minimize these kinds of errors in students' work, teachers need to provide opportunities for students to learn addition in the context of real-life situations.

Addition is generally used in the following four situations. Students will learn the first two concepts in the first grade. The last two will be introduced in the second grade.

1. *Adding-on concept.* To find the final quantity of a group after additional members have been added to the original group. For example: *There are 5 fish in a tank. John put 2 more fish in the tank. How many fish are in the tank altogether?*
2. *Combining concept.* To find the total quantity of two simultaneously existing groups by combining them together. For example: *There are 5 children playing on a slide. There are 2 children playing on the swings. How many children are playing on the playground altogether?*
3. *Comparing concept.* To find the quantity of a group by stating how many more objects of that group exist compared with the number of objects in a reference group of a known quantity. For example: *There is a pile of red and white balls. There are 5 red balls in the pile. There are 2 more white balls than red balls. How many white balls are there?*

4. *Reverse subtracting concept.* To find the original quantity by adding together the final quantity and the quantity that was subtracted. For example: *John used $5 to buy a toy. He has $2 left. How much money did John have in the beginning?*

Important Points for Teaching Addition

When teaching addition, teachers must keep the following points in mind:

- Students need to be able to count accurately before addition is introduced.

- It is important for students to develop the concept of quantity and to use concrete situations when thinking of addition. It is, therefore, essential to use real-life examples (e.g., concrete pictures or word problems that depict the first two types of addition concepts listed in the preceding section and that are related to the students' daily experience) and to use manipulatives to facilitate students' understanding of addition. Students should learn to use terms such as *more, combine, join together, altogether, put together,* and *in total* when discussing addition problems.

- It is also important for students to learn how to represent a concrete situation in abstract terms (i.e., in equation form). By the end of this unit, students should be able to explain how to solve the addition problem and write an equation by using symbols such as + and = to represent the situation.

- Students should learn to express their thinking process from concrete to abstract when solving the problems. They can proceed in the following way:

   Problem: *There are 3 fish in a tank. John put 1 more fish in the tank. How many fish are in the tank altogether?*

   Thinking processes to solve the problem follow:

   1. In the beginning there are 3 fish, and 1 more is added, making a total of 4 fish altogether.
   2. 3 fish and/plus 1 fish are 4 fish.
   3. 3 and/plus 1 are 4.
   4. $3 + 1 = 4$.

   These processes of instructions illustrate the adding-on concept of addition, an increase in group size.

Important Aspects of Equations

In this unit students are exposed for the first time to mathematical equations. It is very important to have students understand what the equations represent. Students should learn that an equation represents not only the numerical calculation but also a simpler way of stating a situation by using symbols instead of words. Teachers should let students explain the equations by using pictures, making up their own story problems, or drawing pictures to correspond to a particular equation. The emphasis on verbal expression of mathematical equations is an important way to ensure that the students thoroughly understand them.

When writing equations, students must know the following points about the components of equations:

1. Equations are composed of symbols that represent *quantities* or *objects,* such as numbers.

2. Equations are composed of symbols that represent the *mathematical operations* to be performed, such as $+, -, \times, \div$.

3. Equations are composed of symbols that represent the *equality or inequality* of single terms or of terms that involve some kind of mathematical operation. Such symbols include $=, <, >$.

A complete equation has to have all three components. Make sure that students do not write incomplete equations, such as $3 + 2 =$ or $3 2 = 5$.

Addition Involving Zero

Students have already learned that *no quantity* is represented by zero. Using this fact, students will learn how to calculate equations that involve zero (e.g., $2 + 0 = 2$, $0 + 3 = 3$, and $0 + 0 = 0$). It may not be easy for students to understand the meaning of these equations if verbal explanations alone are used. Therefore, it is important to have students engage in activities, such as the "toss-a-ball-in-the-box" game, that will show them how to make the equation correspond with actual events. As the students play the "toss-a-ball-in-the-box" game and try to figure out their total score after two trials, they will have an opportunity to understand the setting for addition of zero; for example, $2 + 0$.

Make sure that students understand that the end product (sum) does not increase when zero is added.

Lesson One   Introduction to Addition—
The Adding-on Concept

Lesson Goals
1. To introduce the concept of addition in which additional members are added to a main group
2. To understand how to express the idea of addition with an equation

Materials

Teacher	Students
• Poster (fish tank)	• Workbook (5.1, 5.2, 5.3)
• Poster (bird)	• Homework (5-1)
• Round magnets	

Lesson Plan

Activity One

Introduce the idea of adding objects to a main group and then finding the resulting quantity.

Steps

1. Put the fish-tank poster on the board. Have the students make up stories while looking at the picture on the board and talk freely about the scene. After the students have offered their ideas, summarize what they have said.
2. Make sure the students understand that one fish is being added to an original group of three fish. Finally, have them understand that the main question is, How many fish are there altogether?

Points to Be Aware Of

Always restate the students' relevant comments in a clear, concise way, leading the students to build up the understanding of the adding-on concept.

Activity Two

Use manipulatives to reinforce the adding-on concept of addition.

Steps

1. Now replace the fish with magnets. Begin with the three fish in the tank. Tell the class, *"First there are 3 fish. I will use 1 magnet to represent 1 fish."* Place one magnet on each one of the three fish in the picture. Then move the magnets underneath the picture.

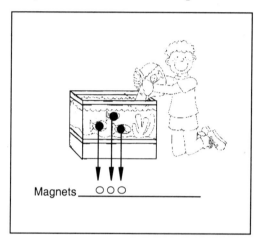

2. *"Then, 1 fish was added."* Place another magnet on the fish being added and then move that magnet underneath the picture.

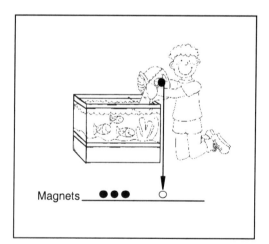

3. *"How many fish are there altogether? Let's count the magnets and find out."* Count as a class, *"1, 2, 3, 4; there are 4 fish altogether."*
4. Repeat the problem straight through aloud. *"First there were 3 fish."* Put the magnets on each of the fish and then move the three magnets down. *"Then 1 fish was added."* Put one magnet on the fish and then move it down. *"There are 1, 2, 3, 4 (counting the magnets), 4 fish altogether."*

Activity Three

Understand how to express the idea of addition with an equation (3 + 1 = 4).

Steps

1. Tell the class: *"We can write what happened in this picture by using numbers. Here is how to write this problem. There are 3 fish in the beginning, so we write a 3."* (Write the 3 on the board.) *"We are adding some fish, so we write the plus sign."* (Write a + next to the 3.) *"How many fish did we add? We added 1 fish. So we write a 1 on the other side of the plus sign."*
2. Go over this part of the equation again if the students seem confused. *"Now, the question asked is, How many fish do we have altogether? To show the processes we are using to figure out how many fish there are altogether, we write an equal sign."* (Write an = sign.) *"Then, we write the total number of fish we have altogether, 4."* (Write a 4.) *"This is an equation. It tells us how many we have to start with, how many were added, and how many we have altogether at the end."*
3. Call on a couple of students and ask them to write the equation and explain to the class the meanings that go with it. They can talk about each part of the equation as they write it.
4. Have the students turn to Workbook page 5.1 and write the equation.[1]

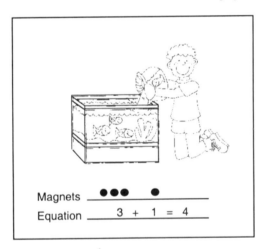

Points to Be Aware Of

It is important to spend time teaching students how to read and express equations. Make sure that the students understand that an equation, through the use of numbers and symbols, can concisely represent the meaning of the situations.

---

[1] The pages for the workbook cited in this lesson are not included in this appendix.

Activity Four

Review the adding-on concept and introduce the drawing-circles method.

Steps

1. Display the bird poster. Ask the whole class to discuss what is happening in the picture.
2. Review briefly the entire procedure (which is the same as that for Activity Three) with the entire class and then call on several students to go through the process:
   - Place magnets on the first group of birds and move the magnets down. Place magnets on the group being added and move those magnets down. (Move the three magnets into the group of four magnets.)
   - Count the total number of magnets aloud.
   - Ask the student, as the student moves the magnets, to describe what he or she is doing and what it means.
3. Ask a different student to come up and write the equation on the board. Have the student explain the meaning of each component of the equation. Have other students elaborate on the meaning of the equation if it is not clearly explained.
4. Display Workbook page 5.2 on the overhead projector. Explain to the students that this time, instead of using magnets, they will draw circles to represent the magnets. (They can use this method when no magnets are available.) *"First there are 4 birds on the tree branch, so we draw 4 circles to represent the birds on the branch. Then 3 birds fly in, so we draw another 3 circles to represent the birds flying in."*

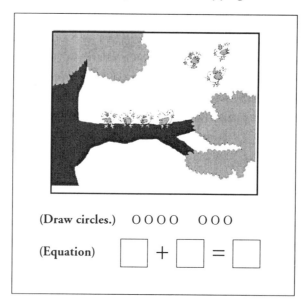

Activity Four (Continued)

5. *"Now, how many birds do we have altogether?"* Have the entire class count and call out the answer.
6. Have the students turn to their Workbook page 5.2. Give the students time to draw their circles and fill in the blanks in the equation part. As the students work, walk around the room to see how they are doing. When they have finished, have two students with different answers come up to the board and write their equations. Have each student explain how he or she came up with the equation. Discuss these equations with the class.
7. Have the students turn to Workbook page 5.3 and work on one problem at a time. After they have finished each problem, ask a student to come to the front of the room, talk about the problem, and write his or her equation for it. Then discuss this equation and the answer. Ask whether any student has any different answer or different explanation to that problem.

Summarize the Lesson

Summarize the adding-on concept and the representation of an equation for the lesson.

Appendix C
Middle School Sample Lesson Plan: Compound Interest

Interest and compound interest is a seventh grade mathematics topic that can be applied in everyday life. Studying compound interest equips students to make real-world decisions about monetary matters. For example, they can more easily understand the costs of "maxing out" credit cards compared with other methods of borrowing money.

Introduction

The first part of this lesson plan gives a detailed discussion of the basic material for the development of that part of the topic appropriate for the seventh grade. The second part, "The Next Steps," develops an application, "doubling time," which is also appropriate for the seventh grade. Also presented in this section are two closely linked topics: some exercises in mathematical reasoning based on the discussion of compound interest that can prepare the students for the properties of exponential functions and a discussion of search procedures that are applicable in a large number of real-world situations. All of this material is presented in the form of actual lessons.

The third part, "Applications of Compound Interest to Loans, Savings, and Present Value of Money," continues the discussion to more advanced topics and is suitable for inclusion in Algebra II. The discussion now focuses on the mathematics itself because it is expected that although the teachers will prepare the detailed lesson plans themselves, they will need a source for the material. These subjects might be suitable for advanced students in the seventh grade. But more likely, the hope is that this material can be covered in the second year of algebra. In any case, it is hoped that most students will be able to see and understand this material before they graduate from high school because it provides the basis for sound decisions involving money management, something that all students need and that the schools should provide.

Additionally, as is the case with most good mathematics, the study of this topic leads naturally to further topics that are quite important for students intending to continue with mathematics. For example, the partial summands of the following geometric series, which first appear in studying compound interest, will appear again and again in algebra and calculus:

$$1 + R + R^2 + R^3 + \cdots + R^n = \frac{R^{n+1} - 1}{R - 1}$$

The natural question of what happens when the time between compounding shrinks to zero leads naturally to the number $e = 2.71828182\ldots$ and, more generally, to the exponential function.

Many teachers of multiple subjects in kindergarten through grade seven may not have the mathematical background to present the topic of compound interest. Consequently, the lesson plans in this appendix carefully lay out the mathematical details of the subject so that the teacher can convey them step-by-step to the students. Because it is critical that the mathematics be understood and presented carefully, a lack of space limits the discussion of pedagogical issues. However, the raw data for filling in the class discussions and assignments are readily available for a topic like this. Numerous advertisements in newspapers and on radio and television offer to lend money at various interest rates and according to certain conditions.

Before the teacher attempts to cover compound interest, students should be able to:

- Perform the basic operations of arithmetic, decimals, and percents; make conversions between fractions, decimals, and percents; and do computations with decimals, percents, and fractions. (Students should be able to work with percents both below and over 100 percent. In the more advanced parts of the subject, multiplying and dividing fractions become important.)
- Raise to powers ($1.03^2 = 1.0609$).
- Apply the rule of exponents $A^m A^n = A^{m+n}$ to some of the more advanced applications.
- Work with the distributive law, as shown below:

$(A \times M) + (B \times M) = (A + B) \times M$

(Be able to translate $P + 0.03 \times P$ into $(1 + 0.03) \times P$.)

- Work with the associative rule, as shown below:

$A(BC) = (AB)C$ and $1.03 \times (1.03 \times M) = 1.03^2 M$

In addition, it would be helpful if the students have been shown the geometric series and geometric sequences. (This topic is not really a prerequisite because the study of savings accounts, present value of money, and mortgage loans can introduce students to geometric series and sequences.)

Beginning the Discussion of Compound Interest

These lessons should begin with a review of percents and simple interest. Previously, students may have experienced vague discussions of percents; for example, 20% is treated as the fraction 1/5. But at this point it is important to emphasize that percents are applied to numbers. For example, one really talks about 20% of 50, which is 10. (In other words, n-percent [#%] is really a function taking numbers to new numbers.)

Next we should review the decimal equivalent of percent. The decimal equivalent of $R\%$ is the decimal number $R/100$, and we write it $D(R)$ when we want to be very clear. $R/100 = D(R)$. This number is multiplied to get the percent of a number; therefore, 1.5% of 130 is:

$$\frac{1.5}{100} \times 130 = 0.015 \times 130 = 1.95$$

$$\boxed{D(R) = \frac{R}{100}}$$

Decimal Equivalent of R %

Sample Problems

Find the following numbers:

1. 10% of 30, 35, 50
2. 20% of 40, 60, 100
3. 1% of 25, 250
4. 2% of 150, 300

With this preliminary out of the way, simple percent applications can be reviewed next. One application is simple discounts, a topic that has a number of applications. For example, in a store one often sees signs showing 10% reductions on all items. This reduction means that the true price of an item will be the original price reduced by 1/10 of the price, or 90% of the price.

Sample Problem

All items in a store are marked down 15%. How much will a ball marked $15 cost?

Another example is a simple interest problem that gives an interest rate and asks the interest to be paid back along with the principal. One gives money to a friend with the agreement that the friend will return the money plus a certain percent; for example, 10%. In which case, if you lend your friend $10, you will get back $11.

Interest problems become more complex when time is introduced as a variable; for example, the idea that an interest rate applies for a particular period of time. However, if the actual time period is greater or less, the amount of interest to be paid will be proportionally more or less.

Sample Problem

If I borrow $200 at a 10% yearly simple interest rate, how much will I owe after a year? How much will I owe after six months?

This type of problem requires the student to understand that computing 10% of $200 tells how much interest needs to be paid after a year; but if the loan is paid in less than a year, the interest will be less.

To determine the interest to be paid after six months, the student must know that six months is half a year. The interest then would be only half of 10%, which is 5%.

Special vocabulary used in the business world in dealing with interest must be taught:

Annual. Yearly, or once a year.

Semiannual or biannual. Twice a year, or every six months.

Quarterly. Four times a year, or every three months.

Interest rate. The percentage that is added on. Thus a 6% interest rate without any further qualifiers means that 6% of your money is added to the money you have at the end of a year.

Interest. This is the *amount* of money that is added on. For example, if you had $100 and the rate of interest was 6%, then, at the end of the year, you would have $0.06 \times 100 = \$6$ in *interest*.

It should now be explained, using real-world examples, that the interest rate advertised by a bank for its savings account, for example, 6% compounded quarterly, really means that at the end of each three-month period, you are given 1.5% interest on the money in the bank during that three-month period. The formula for the interest actually given at the end of the period is:

$$\frac{\text{Total interest for the year}}{\text{Number of periods in year}} = \text{Interest at the end of each period}$$

In the example above, we have $6/4 = 3/2 = 1.5$, which is the interest at the end of each period. Ample practice with the questions of how many times during the year interest is added and how much interest is added to the account each time would be provided. Students' understanding of this concept is obviously critical.

Sample Problems

1. Suppose a bank advertised a 6% interest rate compounded semiannually. How many times during the year is interest added to the account? How much interest is added to the account each time?

2. Suppose a bank advertised a 6% interest rate compounded monthly. How many times during the year is interest added to the account? How much interest is added to the account each time?

3. The problem is the same, except that the bank compounds interest weekly.

At this point the original word formula that follows might be replaced by an algebraic formula:

$$\frac{\text{Total interest for the year}}{\text{Number of periods in year}}$$

Let I_p be the interest for each period. Let I be the total (or annual) interest. Let N be the number of periods (in the year). Then:

$$I_p = \frac{I}{N}$$

It can be emphasized that the two ways of writing this fact are the same. The second way is merely a shorthand for the first. But a student has to remember what I_p, I, and N mean.

Teachers should provide students with ample practice in translating the original word formula into the algebraic formula and in determining what the interest rate per period is. They cannot take for granted that students will be able to do this translation.

The next step would be to write the formula as a decimal equivalent instead of as a percent (with the letter D representing *Decimal*) so that the decimal equivalent $D(I_p)$ can be written as follows:

$$\boxed{D(I_p) = \frac{D(I)}{N}}$$

Decimal Equivalent of I_p

At this point the teacher must provide students with practice again. It is very important that the students understand that in calculating the amount of interest for each period, they have to use the *decimal equivalent* of percent. A common error that indicates a lack of understanding is that the student, when doing calculations, writes a monthly interest of 1% as 1 instead of as 0.01.

Note: It might be well worthwhile to alert the students to the various rules that banks have for determining the amount on which interest is calculated in an account. For example, interest is almost never calculated on the balance in the account at the end of the period but, instead, is usually calculated on some kind of average daily balance. Depending on the rules, students should discuss the best strategies for withdrawing and depositing money during an interest period. But the teacher should emphasize that in every case, if money is left in the account for the entire interest period, and not touched, then the bank will pay interest on the entire amount.

Introducing the Calculation of Compound Interest—The Long Way

The process of what happens during the compounding of interest should be illustrated for students. They must understand that compound interest earns more than simple interest.

They must understand that each time the interest for a period is calculated, the amount is determined by the amount the depositor started with and all the interest that has been added to the original amount.

Suppose that $10,000 is deposited in a bank, and each month the interest is compounded. The interest rate for a year is 6%; therefore, the interest rate for each one-month period is 0.5%. The table shown below illustrates compound interest. Early in January $10,000 is deposited, and 0.5% interest is added at the end of January. The interest is figured by multiplying $0.005 \times \$10,000$, which is $50 interest:

Month	Money in bank at beginning of month	Interest added during month	Money in bank at end of month
January	$10,000	$50.00	$10,050

At the beginning of February, $10,050 is in the account. The interest is computed at $0.005 \times \$10,050$. The interest for this period is $50.25. The account earned a little more interest than it did in January because more money was in the account at the beginning of February.

Month	Money in bank at beginning of month	Interest added during month	Money in bank at end of month
January	$10,000	$50.00	$10,050
February	$10,050	$50.25	$10,100.25

At the beginning of March, $10,100.25 is in the account. Computing the interest again gives $10,100.25 \times 0.005 = \$50.50$. Again this amount is more than the interest received in January or February because at the beginning of March, more is in the account than there was in February.

Month	Money in bank at beginning of month	Interest added during month	Money in bank at end of month
January	$10,000	$50.00	$10,050
February	$10,050	$50.25	$10,100.25
March	$10,100.25	$50.50	$10,150.75

Students are guided in completing the table for the entire year. Then the amount of interest compounded is compared with the amount of simple interest earned during a year.

Students practice filling in tables, given different principal, interest rates, and time periods.

Introducing the Formula

Students must understand that the formula for finding the compounded interest is really a fast way to compute what can be found out by making many individual calculations. Here is the formula:

$(1 + D(I/N))^m \times S$

$D(I/N)$ gives the interest rate per period *in decimal form*. The letter m is the total number of periods that interest will be compounded. The letter S is the amount started with.

Students can be shown the process through which the formula comes about by setting up a table in which the original amount is written as a letter instead of as an amount. In the example shown below, the interest rate is 12% annually, and interest is compounded quarterly so that the interest rate for each period is 3% and the amount of interest is determined by multiplying by 0.03.

The letter S represents the amount in the bank at the start.

(Note that the participation of students in this activity depends on their knowledge of all the component skills involved.)

Month	Money in bank at beginning of month	Interest added during month	Money in bank at end of month
First	S	$0.03S$	$S + 0.03S$ or $(1.03)S$

The teacher has to be sure that students understand how and why $S + 0.03S$ can be rewritten as $(1.03)S$. (This is an application of the distributive rule.) The teacher shows that:

$$S + 0.03S = 1 \times S + 0.03 \times S$$
$$= (1 + 0.03) \times S$$

The calculation for the second period is made as follows:

Month	Money in bank at beginning of month	Interest added during month	Money in bank at end of month
First	S	$0.03S$	$(1.03)S$
Second	$(1.03)S$	$0.03 \times (1.03)S$	$(1.03) \times (1.03)S$ $= (1.03)^2 S$

The teacher shows that:

$$1.03 \times S + 0.03 (1.03) \times S = 1 \times (1.03) \times S + 0.03 \times (1.03) \times S$$
$$= (1 + 0.03) \times (1.03) \times S$$
$$= 1.03 \times (1.03) \times S$$
$$= (1.03)^2 \times S$$

For emphasis the calculation for the third period is made as follows:

Month	Money in bank at beginning of month	Interest added during month	Money in bank at end of month
First	S	$0.03S$	$(1.03)S$
Second	$(1.03)S$	$0.03 \times (1.03)S$	$(1.03)^2 S$
Third	$(1.03)^2 S$	$0.03 \times (1.03)^2 S$	$(1.03)^3 S$

The teacher shows that:

$$(1.03)^2 \times S + (0.03) \times (1.03)^2 \times S = 1 \times (1.03)^2 \times S + (0.03) \times (1.03)^2 \times S$$
$$= 1.03 \times (1.03)^2 \times S$$
$$= (1.03)^2 \times S$$

At this point it should be clear to some students that a pattern for predicting the amount is emerging. At the end of the mth period, the ending amount can be determined by the amount at the beginning of the mth period:

Amount at end of mth period = (1.03) × amount at beginning of mth period

To determine the amount at the end of the fourth period, we multiply the amount from the end of the third period by 1.03 so that we get:

$$(1.03) \times (1.03)3 \times S = (1.03)^4 \times S$$
$$= 1.125508 \times S$$

If the original amount was $10,000, after a year with an interest rate of 12% compounding quarterly, the account would contain $11,255.08 compared with $11,200 if straight interest was earned.

Note that the students must be able to multiply 1.125508 × 10,000 in order to do this problem. Again, let us stress that without knowledge of all the components, such as multiplication with decimals and rounding off, which are needed in many kinds of calculations, students will not be able to understand or even do these problems.

The work described above sets the stage for introducing the general formula for computing compound interest:

$$\boxed{\left(1 + D\left(\frac{I}{N}\right)\right)^m S.}$$

Main Formula

Students must be able to explain what each letter and component of the formula represents:

- $D(I)/N = D(I/N)$ is the interest rate written as a decimal to be applied during each period and calculated by dividing the annual interest rate by the number of periods each year.
- The letter m is the total number of periods during which interest will be compounded.
- The letter S is the amount at the start.

It is useful to introduce the terminology that *the multiplier* in the main formula for compound interest is the decimal number $(1 + D(I/N))$.

The teacher provides sets of questions to test the students' ability to put numbers in the formula and demonstrate understanding. For example, the students are given the problem shown below: they are to substitute for letters in the formula, do the calculations, and answer questions, such as the following:

1. John deposits $1,000 for a period of one year. How much money will he have in his account at the end of 9 months if the bank is offering an interest rate of 5% compounded monthly?

    a. What is the annual interest rate?
    b. By what number do you divide the annual interest rate to determine the interest rate for each period?
    c. For how many periods will interest be given?

2. Find the multipliers when:

    a. The interest rate is 8% compounded biannually. (1.04)
    b. The interest rate is 6% compounded monthly. (1.005)
    c. The interest rate is 7% compounded quarterly. (1.0175)

Next, students are expected to work a range of applications.

Sample Problems

1. Suppose we deposit $1,000 at the beginning of an interest period and leave the money in the account for a year. Suppose that the interest rate is 12%, compounded quarterly. How much will we have at the end of one year? (First set the problem up. Then use your calculator to solve it.)

2. Suppose we deposit the same $1,000 at the beginning of an interest period and leave the money in the account for a year. But now suppose that the interest rate is 12%, compounded monthly. How much will we have at the end of one year? (First set up the problem. Then use your calculator to solve it.)

3. If John deposits $1,000 for a period of one year, how much money will he have in his account at the end of the year if the bank is offering an interest rate of 5% compounded semiannually? (First set up the problem. Then use your calculator to solve it.)

4. If John deposits $1,000 for a period of one year, how much money will he have in his account at the end of the year if the bank is offering an interest rate of 5% compounded monthly?

Use of tables. Students may also complete tables, as shown in the following example, in which the initial amount is $1,000 and the interest rate is 10%.

Table 1
Accumulation of Compound Interest
(In dollars)

Periods	1 year	2 years	5 years	10 years	25 years
$N = 1$	1,100	1,210	1,610.512	2,593.74	10,834.71
$N = 2$	1,102.50	1,215.51	1,628.89	2,653.30	11,469.40
$N = 3$	1,103.36	1,217.40	1,635.26	2,674.06	11,693.05
$N = 6$	1,104.28	1,219.44	1,642.10	2,596.50	11,939.92
$N = 12$	1,104.73	1,220.44	1,645.47	2,707.58	12,062.93

Sample Problems

1. Create a similar table if the interest rate is 8%; N indicates the number of periods in a year, $N = 2, 4, 6$, and 12; and the lengths of time are as shown above, 1 year, 2 years, 5 years, 10 years, and 25 years.

2. Create a table for $N = 2, 4, 6$, and 12 with time lengths of 1 year, 2 years, 5 years, 10 years, and 25 years if the interest rate is 6%.

Use of graphs. Students may also create graphs, as shown in figure 1, in which growth under compound interest starts slowly and increases rapidly. The following formula and graph demonstrate growth under compound interest:

$$\left(1 + \frac{I}{N}\right)^m S$$

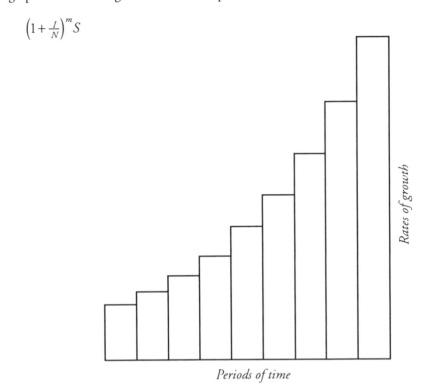

Figure 1. Growth Under Compound Interest

As students work with compounding problems, the teacher may present questions involving applications during discussions:

1. Does compounding more frequently result in a larger sum at the end of a time period?
2. What happens to the difference between simple (annual) interest ($N = 1$) and more frequently compounded interest rates as the amount of time increases?
3. Would it still be better to have the interest compounded more frequently if the interest rate is very high?

The Next Steps

At this point students should have mastered the basic idea of compound interest. The next topic covers a basic application for figuring out the length of time needed for money in an account to double when the interest is compounded.

In further applications what happens when money is withdrawn from or added to an account will be discussed and, finally, all these topics will be combined in the three final applications: savings plans, present value of money, and bank loans. These last applications are more advanced technically, but they are extremely important as applications of mathematics that almost all students will need to understand once they enter the workplace. Consequently, if it is at all possible, these topics should at least be discussed in the classroom.

Doubling Time

One of the more interesting applications of compound interest for students and teachers is the question of how long it would take to double the money deposited in an interest-bearing account in which the interest is compounded.

Working with questions about doubling time not only is interesting in itself, but also can serve several other purposes. First, it provides more practice to help students get a better feel for the growth rate of money when the interest is being compounded. Second, it provides practice with a systematic search procedure. There are innumerable cases in which no formulas exist for finding out something. (To give a formula here would require the use of logarithms.) Therefore, the answer has to be found by using a case-by-case analysis. Doing this analysis efficiently is very important and a basic illustration of the methods of thought that mathematics should instill in students.

The question to consider is, how long will it take to double your money?

A teaching sequence might begin by having students figure out how long it would take to double the original amount in a bank account, given simple interest. The students review this concept: when simple interest is applied, the original amount is doubled when the interest rate is 100%. Teachers may present a case of simple interest of 10% per year, which is not compounded at all, so that when the money is withdrawn after N years, the depositor will get the money back plus interest of $N \times 10\%$ (the number of periods of time multiplied by 10% each period). Given that each period is a year, the interest rate would be 100% after ten years. The students can calculate that a one-dollar investment would become two dollars in ten years, given an annual interest rate of 10% that is not compounded at all.

Next, the teacher helps the students in determining what will happen if the interest rate remains 10%, but the rate is compounded 4 times annually. The students are again asked to find out how long it would take to at least double the money in the account.

Since we know that the amount of money in the account after m interest periods is $(1 + D(I/N))^m S$, the goal of the exercise is to determine the smallest m that makes the inequality shown below true. Here S equals the original amount deposited, I the annual interest rate, N the interest periods annually, and m the total number of interest periods:

$$(1 + 0.025)^m S \geq 2S \text{ or } (1.025)^m S \geq 2S$$

(Of course, in this example it is assumed that $S = \$1$.) As before, the students determine the interest rate per period by dividing the 10% annual rate by 4 and by writing the interest rate per period as a decimal. They have to find the first m (the smallest number of periods needed) to get an answer of 2 or more: $(1 + 0.025)^m \times S \geq 2 \times S$. (The calculation $2 \times S$ represents a doubling of the amount the depositor started with.)

$$(1 + 0.025)^m S \geq 2S \text{ or } (1.025)^m S \geq 2S$$

At this point the teacher or the class should observe that S does not really matter here. What matters is whether $(1.025)^m \geq 2$ or not. If it is ≥ 2, then the result will be at least $2S$. If not, then the result will not be at least $2S$. So *the real problem is to find the first m so that* $(1.025)^m \geq 2$.

Depending on the tools available, this calculation can be done in various ways. First, if a computer is available, it can be programmed to calculate successive powers of (1.025) until the first one that is greater than or equal to 2 is found.

```
     M = 1
     Mult = 1.025
10   V = (Mult)^M
     IF V ≥ 2 PRINT M: END
     ELSE M = M + 1: GOTO 10
     END
```

Otherwise, a method of searching can be used.[1] The teacher might ask the students to begin their search by calculating how much money would be in the account after 10 years. (Trying 10 years first is reasonable because more money is earned from compound interest than from simple interest and simple interest doubles the money in the account in 10 years.)

---

[1] Mathematically speaking, the students will learn more from the process of systematically searching for the answer than from using the computer, but both methods are valuable and teach important skills.

The teacher helps students determine that for 10 years the *m* is 40 (interest is compounded quarterly for 10 years). (Note that the teacher must assess whether students can readily determine the number of periods, *m*, when they are given the number of times interest is compounded and a total time period. This topic has already been covered in the compound interest section, but seeing the same concept in a new situation helps students to solidify their understanding.)

The students substitute 40 for *m* and then perform the calculation on their calculator. The answer is:

$$(1.025)^{40} = 2.6850638\ldots$$

which is considerably greater than 2, so the time will be much less than 10 years. The teacher then leads the students through checking for shorter time periods until they find that after 7 years, or 28 interest periods, the result is $(1.025)^{28} = 1.996495$. After 29 interest periods, or 7 years and 3 months (7.25 years), the result is 2.0464074. Consequently, the doubling period will be 7 years and 3 months, which is quite a bit less than 10 years.

The process here might be done as shown below (with calculators the students do the calculation for 8, 16, 24, and 32 periods):

$$(1.025)^{8} = 1.2184\ldots$$
$$(1.025)^{16} = 1.4845\ldots$$
$$(1.025)^{24} = 1.80872\ldots$$
$$(1.025)^{32} = 2.203756\ldots$$

(It should probably be mentioned *why* the above numbers were selected. Here, 8 periods is 2 years. So we are really checking what happens at the end of 2 years, 4 years, 6 years, and 8 years because 32 periods is 8 years.)

Doubling will have occurred sometime between the end of the *sixth* year and the end of the *eighth*. Calculators can again be used to close in on the exact time. A direct way of doing this calculation is to go by 2s to check $m = 26, 28, 30$:

$$(1.025)^{26} = 1.90029\ldots$$
$$(1.025)^{28} = 1.99649\ldots$$
$$(1.025)^{30} = 2.09756\ldots$$

and, since $(1.025)^{30} > 2$, the only question is whether this is the first *m* for which this happens or whether $(1.025)^{29} > 2$ as well. So we check:

$$(1.025)^{29} = 2.046\ldots$$

and the *first time* doubling occurs is $m = 29$; therefore, doubling occurs after 7 years and 3 months.

We should summarize. The rule for finding how long it will take to double your money is to find the first *m* so that:

$$(1 + D(I_p))^m \geq 2$$

and then to multiply *m* by the time period.

The teacher can write this rule down and ask the students to explain it. They should understand that since the amount after k periods is $(1 + D(I_p))^k S$, then to have at least $2S$, it must be the case that $(1 + D(I_p))^m \geq 2$.

Teachers can work with students on a variety of applications. They can work on making a table that tells doubling time, given particular interest rates and compounding periods. Teachers can structure the exercises so that students will know when to try longer or shorter time periods and can explain why their answers are correct. ("I know that the doubling period was more than 7 years because we ended up with an amount that was less than 2 when we calculated for the periods in 7 years, but the amount was greater than 2 when we calculated for the 29th period, which is 7 years and 3 months.")

Doubling Times

Rate (Percent)	$N=1$ (In years)	$N=2$ (In years)	$N=4$ (In years)	$N=12$ (In years)
3	24	23.5	23.25	23.17
6	12	12	11.75	11.6
7	11	10.5	10	10
9	9	8	8	7.75
10	8	7.5	7.25	7

Students can also be given problems with actual amounts:

Jason deposited $2,500 in the bank. The account will earn 5% annually and be compounded quarterly. How long will it take for Jason to have at least $5,000 in that account?

Further Developments from Doubling Time

If time is available, two topics can be covered that represent opportunities for students to practice the critical skills of abstraction in a way that *manifestly* increases their ability to deal with real-world situations. The first is to develop the discussion described below on what happens if the doubling period is known exactly. What happens after two doubling periods, three doubling periods, one-half of a doubling period? The second topic is a discussion of how to do a search procedure.

Exactly known doubling period. Remember that the main formula is $\left(1 + \dfrac{D(I)}{N}\right)^m S$.

Let us suppose that $N = 4$ and the doubling time is *exactly* 6 years.

1. How much will your money increase in exactly 12 years?

   Note that $\left(1 + \dfrac{D(I)}{4}\right)^{24} = 2$ since the assumption is that the principal exactly doubled after 6 years, or 24 interest periods. In 12 years there will be 48 periods, which gives:

$$\left(1 + \frac{D(I)}{4}\right)^{48} = \left(1 + \frac{D(I)}{4}\right)^{24+24}$$

$$= \left(1 + \frac{D(I)}{4}\right)^{24} \times \left(1 + \frac{D(I)}{4}\right)^{24}$$

$$= 2 \times 2$$

$$= 4$$

2. How much will your money increase in exactly three years?

Note again that $\left(1 + \frac{D(I)}{4}\right)^{24} = 2$. The amount of increase for 3 years is thus:

$$\left(1 + \frac{D(I)}{4}\right)^{12} = M. \text{ Note that:}$$

$$M \times M = \left(1 + \frac{D(I)}{4}\right)^{12+12}$$

$$= \left(1 + \frac{D(I)}{4}\right)^{24}$$

$$= M^2$$

Thus $M^2 = 2$, so the amount of increase in 3 years will be $\sqrt{2}$, or 1.41421 . . . .

Sample Problem

Suppose that the doubling period is exactly 6 years and suppose that we deposit $100. How much will we have after 30 years? ($3,200) How much after 36 years? ($6,400) How much after 60 years? ($102,400) How much after 63 years? ($144,815.47)

Search procedure. The second topic is to amplify the discussion of *search procedure* indicated in the original analysis. In real-world situations searching huge databases is a very important and time-consuming task. Efficient procedures had to be developed to do these searches quickly, and such a procedure can be well illustrated here.

We can be very systematic in our search procedure if we search by cutting the interval in half each time the search takes place. We illustrate this method by using a 12% interest rate compounded 12 times each year; therefore, the interest in each time period is 1%. Certainly, the doubling time is less than 100 because 1% is added after each period; and after 100 periods *at least* 100% will be added.

Now, half of 100 is 50, so we try $(1.01)^{50} = 1.6446$ . . . , which is too small. Therefore, the doubling period is greater than 50. But it is also less than 100.

Midway between 50 and 100 is 75. So we try $(1.01)^{75} = 2.10912$ . . . . This number is larger than 2, so m may be less than 75, but it is not larger than 75. It is, however, larger than 50.

Midway between 50 and 75 is 62.5. So we try 63: $(1.01)^{63} = 1.87174\ldots$. This number is too small. So the doubling time is between 63 and 75. Midway between is 69: $(1.01)^{69} = 1.98689\ldots$.

The desired m is greater than 69 and less than or equal to 75, so we try 72, which is midway:

$(1.01)^{72} = 2.04709\ldots$

Now we are really close. So we check 70 and 71 to see which is the *first* m so that $(1.01)^m \geq 2$, as shown below:

$(1.01)^{70} = 2.0067\ldots$

The result is that $m = 70$, and the doubling period is 70 months, or 5 years and 10 months.

The Search Procedure

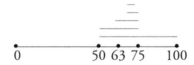

Applications of Compound Interest to Loans, Savings, and Present Value of Money

The previous topics comprise what should be covered in seventh grade. The topics discussed below can be covered in a preliminary way in the seventh grade. The formulas for savings accounts and loans can be given to the students, who can be given exercises to become familiar with the formulas. Of course, for the best students and in the better classes, more can be done, depending on the judgment of the teacher.

However, Algebra II, in which geometric series are studied, is also a very natural place for these topics to be discussed in depth (Algebra II Standards 22.0 and 23.0). They provide wonderful motivation for studying geometric series. And, again, they provide real and vitally useful applications of mathematics that can be used to motivate the students and help them stay focused on the course. Here, more than in the previous part of this appendix, the emphasis is on the mathematics involved.

The two processes of adding and subtracting money from an account that is drawing compound interest are discussed. Included are the basic subjects of bank loans and mortgages, savings accounts, and such sneaky practices as lottery payouts. The discussion starts with seeing the effect of a *single* withdrawal or a *single* deposit over time. This topic is then applied to the general situation.

Withdrawing Money from an Account

Suppose that we have an account that is accumulating compound interest, and we withdraw money during the kth interest period. What happens during further interest periods? Let I be the rate of interest and N the number of periods. Suppose that S was the amount originally in the account and A is the amount that is withdrawn. Then the amount of money in the account during *later interest periods* will be given by the formula:

$$\left(1 + \frac{D(I)}{N}\right)^m S - \left(1 + \frac{D(I)}{N}\right)^{m-k+1} A$$

where m is greater than k, since we are looking at time after the money is withdrawn. This formula means that for $m > k$ we lose not only A but also the interest that A would have earned during the $m - k + 1$ interest periods when it was not in the account.

Figure 2 shows the effect of removing money in the third interest period. At later stages the interest on this money is also removed, as shown in the following formula:

$$\left(1 + \frac{D(I)}{N}\right)^m S - \left(1 + \frac{D(I)}{N}\right)^{m-2} A$$

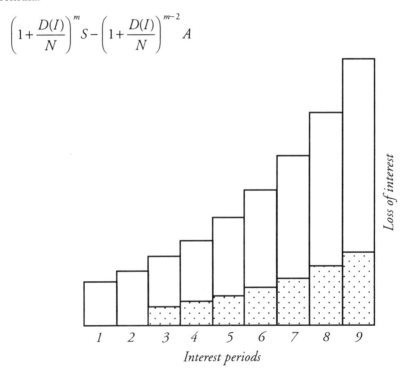

Figure 2. Effects of Withdrawing Money

Sample Problems

1. Set up the formula for determining the amount of money in an account at the end of 4 years. We initially put $1,000 in an account with a 6% interest rate compounded monthly and withdraw $200 at the end of 6 months.

2. Suppose we invest $1,500 in an account with a 7% interest rate compounded quarterly. Suppose that at the end of the first year, we withdraw $400. How much would we have at the end of the third year?

Adding Money to an Account

Adding money to an account that is drawing compound interest at the beginning of the kth interest period does the reverse of withdrawing money. The added money begins drawing interest as well, but since it has been in the bank for less time, it draws less interest. The formula appears as follows:

$$\left(1+\frac{D(I)}{N}\right)^m S + \left(1+\frac{D(I)}{N}\right)^{m-k+1} A$$

The formula shows the total amount after m interest periods if $m \geq k$. The only difference between this formula and the one for withdrawing A is that in this formula the interest for A is *added* instead of subtracted.

The effect of adding money for the third interest period is shown in figure 3:

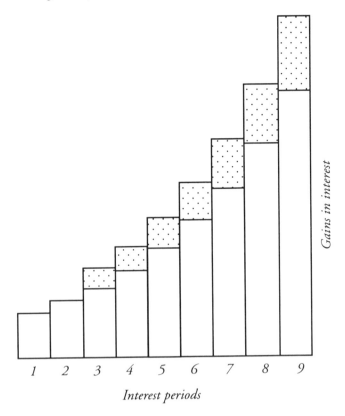

Figure 3. Effects of Adding Money

At later stages the interest on this money is also added, as shown in the following formula:

$$\left(1+\frac{D(I)}{N}\right)^m S + \left(1+\frac{D(I)}{N}\right)^{m-2} A$$

Some sample problems that are similar to the problems for withdrawing money should also be introduced both in class and as homework. The first objective is to practice setting up the formula. Then, after the students understand how to do so, they should work a number of explicit examples.

In the material that follows, the preceding formulas are applied to problems involving savings accounts in which a fixed amount is added every month, the present value of money, and bank loans.

Savings Plans, Present Value of Money, and Bank Loans

The previous discussion gives the basic ideas of compound interest, but this topic can be explored further. For example, when people get loans from a bank, the bank designates the rate of interest, the time needed to pay off the loan, and the amount of the payments. How do banks determine these figures? Understanding this process will help in managing money and provide basic tools for making informed decisions about many kinds of monetary transactions.

Savings plans—saving for college. Often, when a child is born, his or her parents might set up a savings plan to put aside a small amount of money every month to help pay for college when the child grows up. Let us assume that the money is put into a savings account every month and that interest is compounded on the account every month. How much will be in the account at the end of m interest periods? The procedure for determining the amount is as follows:

We start with a sum S.

1. At the end of the interest period, we add an amount A. But the bank adds the interest on S. This interest is called I_1.

    Now we have $(S + I_1) + A$.

2. At the end of the next interest period, we get the interest on $(S + I_1) + A$, and we add A again. The interest is I_2.

    Now we have $(((S + I_1) + A) + I_2) + A$.

3. At the end of this interest period, we have the interest on $(((S + I_1) + A) + I_2) + A$, and we add A again. We write the interest on $(((S + I_1) + A) + I_2) + A$ as I_3.

    Now we have $((((S + I_1) + A) + I_3) + A) + I_2) + A$, and so on.

Here is the explicit formula. After m interest periods the amount in the account will be:

$(1 + D(I_p))^m S + (1 + D(I_p))^{m-1} A$
$\quad + (1 + D(I_p))^{m-2} A + (1 + D(I_p))^{m-3} A + \cdots + (1 + D(I_p))A + A$

This formula is explained by stating that the original amount S gets interest for all m interest periods, the first added A gets interest for $m - 1$ interest periods so that it appears multiplied by $(1 + D(I_p))^{m-1}$, the second added A gets interest for $m - 2$ interest periods so that it appears multiplied by $(1 + D(I_p))^{m-2}$, and so on, to the last A, which does not get any interest yet.

Instead of doing the very long sum above, there is a way of *rewriting* the formula to make calculations easier:

$$\boxed{(1 + D(I_p))^m S + \left(\frac{(1 + D(I_p))^m - 1}{D(I_p)}\right) A}$$

Note that the sum:

$$(1 + D(I_p))^{m-1} A + (1 + D(I_p))^{m-2} A + \cdots + (1 + D(I_p)) A + A$$

in the original formula is replaced by the much simpler expression in the new formula:

$$\left(\frac{(1 + D(I_p))^m - 1}{D(I_p)}\right) A$$

Before explaining this replacement, the teacher gives the students practice in using the simplified formula.

Suppose that every month $40 is put into the account and the interest is 6% compounded monthly. How much will be in the account after 15 years?

Here, $I_p = \frac{6}{12}\% = 0.5\%$ so that $D(I_p) = 0.005$. Also, since we start with just the first payment, the initial amount $S = \$40$. Finally, the number of interest periods is $12 \times 15 = 180$. The formula below shows that the amount after 15 years will be:

$$\$\left((1.005)^{180} \times 40 + \frac{(1.005)^{180} - 1}{0.005} \times 40\right)$$

This amount is calculated to be $11,731.

For information we should also note that we will have put in $40 \times 180 = \$7,200$; therefore, the remaining $4,531 is the amount of interest the money has earned.

Sample Problems

1. Suppose that we start with $150 and at the end of every interest period, we put in $100. How much will be in the account at the end of 5 years if the interest is 6% compounded quarterly?

2. Suppose that we start with $500 and at the end of every interest period, we put in $100. Assume that the interest is compounded quarterly.

   a. Make a table showing how much we will have at the end of 5 years if the interest is 7%, 8%, 9%, and 10%.

   b. Make a table showing how much we will have at the end of 10 years if the interest is 7%, 8%, 9%, and 10%.

c. Make a table showing how much we will have at the end of 15 years if the interest is 7%, 8%, 9%, and 10%.

Explaining the Formula (optional)

What is needed is to explain why the sum:

$$(1 + D(I_p))^{m-1}A + (1 + D(I_p))^{m-2}A + \cdots + (1 + D(I_p))A + A$$

is replaced by the expression:

$$\left(\frac{(1+D(I_p))^m - 1}{D(I_p)}\right) A$$

First, we should take the A common to each term in the sum and, using the distributive law, multiply A once times the simpler sum:

$$(1 + D(I_p))^{m-1}A + (1 + D(I_p))^{m-2}A + \cdots + (1 + D(I_p))A + A$$
$$= [(1 + D(I_p))^{m-1} + (1 + D(I_p))^{m-2} + \cdots + (1 + D(I_p)) + 1] A$$

This approach is still too complicated. To make writing easier and also to focus on the important point in the simplification, the teacher should now replace $(1 + D(I_p))$ with R each time that formula appears in the expression shown previously. So the expression becomes:

$$[R^{m-1} + R^{m-2} + \cdots + R^2 + R + 1] A$$

The next step is to simplify the sum inside the square brackets. It can be phrased in the form of a lemma since this result is very important and will occur often in more advanced classes.

LEMMA: *Let R be any number other than 1, then:*

$$R^m + R^{m-1} + R^{m-2} + \cdots + R^2 + R + 1 = \frac{R^{m+1} - 1}{R - 1}$$

(One may check this equation by multiplying both sides by $R - 1$, which can be done since $R - 1 \neq 0$. Note that $(R - 1) \times V = R \times V - V$. But R times the sum shown above is obtained by multiplying every term in the sum by R:

$$R^{m+1} + R^m + R^{m-1} + \cdots + R^3 + R^2 + R$$

Then, when the original sum is subtracted, the subtraction can be organized as follows:

$$R^{m+1} + R^m + R^{m-1} + \cdots + R^2 + R$$
$$- R^m - R^{m-1} - \cdots - R^2 - R - 1$$

Subtracting term by term shows that all the interior terms cancel, leaving $R^{m+1} - 1$ as $(R - 1)$ times the sum.

Now, in the original formula $(1 + D(I_p)) = R$, so we substitute:

$$\frac{R^m - 1}{R - 1} = \frac{(1 + D(I_p))^m - 1}{(1 + D(I_p)) - 1}$$

We can note that in the denominator $(1 + D(I_p)) - 1$, both of the ones cancel so that the denominator becomes $D(I_p)$ and, finally, we have written the sum in square brackets as:

$$\frac{(1 + D(I_p))^m - 1}{D(I_p)}$$

When we multiply by A, we get the simplified formula for the amount in the savings account after m interest periods.

Present value of money. False claims of lottery payments is a scam that happens all the time. Suppose that you win one million dollars in the lottery. According to the rules this amount will be paid out to you in equal payments spread out over 20 years. How much will each payment be?

$$\frac{\$1,000,000}{20} = \$50,000$$

But how much will your million dollars really cost the lottery? What it might do is to put a certain amount of money in the bank where it will earn interest and pay you $50,000 each year from that account for 20 years until all the money in the account is used up. In this case, how much would the lottery have to put into the bank to pay you the $1,000,000?

The answer may surprise you. *It will turn out that under relatively conservative assumptions the lottery has to put only $529,700.71 into the account.* This is *the present value* of the $1,000,000.

We can figure out *the present value* in this way. Let us suppose that the bank pays 7% interest yearly, the money is put in, and you get your first payment one year later. Then at the end of one year, if S is the amount deposited in the bank, the account will have the following amount:

$$(1.07) \times S - 50{,}000$$

Then, at the end of the second year, it will have:

$$(1.07) \times ((1.07) \times S - 50{,}000) - 50{,}000$$

This equation can be written as:

$$(1.07)^2 \times S - ((1.07) + 1) \times 50{,}000$$

At the end of the third year, the account will have:

$$(1.07) \times ((1.07)^2 \times S - ((1.07) + 1) \times 50{,}000) - 50{,}000$$

This equation can be written as:

$$(1.07)^3 \times S - ((1.07)^2 + (1.07) + 1) \times 50{,}000$$

The procedure shows that the amount in the account after m years will be:

$$(1.07)^m \times S - ((1.07)^{m-1} + (1.07)^{m-2} + \cdots (1.07) + 1) \times 50{,}000$$

When $m = 20$, we would expect that the account should be used up, so we have the equation:

$$(1.07)^{20} \times S - ((1.07)^{19} + (1.07)^{18} + (1.07)^{17} + \cdots + (1.07) + 1) \times 50{,}000 = 0$$

This equation can be rewritten when we use our equation for the sum:

$$R^m + R^{m-1} 1 + \cdots + r^2 + R + 1$$

as shown below:

$$(1.07)^{20} S - \left(\frac{1.07^{20} - 1}{0.07}\right) \times 50,000 = 0$$

Dividing both sides by $(1.07)^{20}$ shows that the original amount that the lottery actually paid to give you your $1,000,000 was:

$$\boxed{S = \left(\frac{1}{(1.07)^{20}}\right)\left(\frac{(1.07)^{20} - 1}{0.07}\right) 50,000}$$

By using our calculators, we find that, as we claimed:

$$S = \$529,700.71$$

This amount is the actual payout by the lottery, just a little more than one-half of the claimed amount.

The general formula will then be:

$$\boxed{S = \left(\frac{1}{(1 + D(I_p))^m}\right)\left(\frac{(1 + D(I_p))^m - 1}{D(I_p)}\right) A}$$

I_p is the interest per period, m is the number of periods, S is the present value, and A is the payout for each period.

Sample Problems

1. How much would the lottery have to pay if the interest the bank paid was assumed to be 8%? Answer the same question with a 6% interest rate.

2. How much would the lottery have to pay for a $1,000,000 prize if the payout is over a 25-year-period instead of a 20-year-period and the interest rate is assumed to be 6% per year?

As in the previous section, these problems test students' understanding of the terms in the boxed formula shown above. In each case the student must figure out the terms $D(I_p)$, m, and A in order to calculate S. In this process students use the skills they learned in the previous sections of this appendix, but use them in a new context. This approach solidifies students' understanding.

The cost of borrowing money. The same considerations apply to getting loans from a bank, such as a mortgage or a loan to buy a car or even a bicycle. A bank lends a certain amount of money S and charges a certain % interest, for example, 8%, which is usually compounded monthly.[2] Every month the borrower is required to pay the bank a fixed amount of money, M. These payments will pay off the loan in 15 years or by whatever time has been agreed on.

Suppose that you borrow $2,000 at 9% interest and you want to pay off the loan in 3 years. How much will you have to pay each month in order to do this? How much will the $2,000 really cost you?

The general formula for this topic is:

$$M = \left(\frac{D(I_p)(1 + D(I_p))^m}{(1 + D(I_p))^{m-1}} \right) S$$

In this formula M is the monthly payment, S is the total amount of the loan, I_p is the interest for each period, and m is the total *number* of payments.

As usual, students should practice with the formula. For example, incorporating the previously given assumptions in the formula will show that the monthly payments are $63.60.

Explaining the Formula (optional)

We consider in detail the situation of the three-year bank loan for $2,000 at 9% interest.

At the end of the first month, the bank will add 0.75% of $2,000 as interest, and you will pay M dollars, leaving the amount yet to be paid off as:

$(1.0075) \times 2,000 - M$

At the end of the second month, you will again pay M dollars, and the bank will add 0.75% of the remaining amount; therefore, you will still owe money on:

$(1.0075) \times ((1.0075) \times 2,000 - M) - M$

This formula can be written as:

$(1.0075)^2 \times 2,000 - (1.0075) \times M - M$

At the end of the next period, the bank will add its interest, and you will again pay M dollars. So what will be the remaining unpaid balance?

$(1.0075) \times ((1.0075)^2 \times 2,000 - (1.0075) \times M - M) - M$

This equation can be written as:

$(1.0075)^3 \times 2,000 - (1.0075)^2 M - (1.0075)M - M$

---

[2] With the increased computing power now available, many banks compound the interest on their loans daily. But the loan is still paid in monthly payments. The class can discuss the fact that this practice actually increases the interest the bank charges.

or as:

$$(1.0075)^3 \times 2{,}000 - ((1.0075)^2 + 1.0075) + 1) \times M$$

It seems that we are getting exactly the same equations as we did before. Therefore, after m payments, the amount of money still owed will be:

$$(1.0075)^m \times 2{,}000 - ((1.0075)^{m-1} + (1.0075)^{m-2} + \cdots + (1.0075) + 1) \times M$$

As shown before, this formula can be rewritten as:

$$(1.0075)^m \times 2{,}000 - \left(\frac{(1.0075)^m - 1}{0.0075}\right) \times M$$

Now we suppose that the loan is to be paid off in 3 years (or 36 payments). This means that after 36 payments the total owed will be zero, so we get the equation:

$$(1.0075)^{36} \times 2{,}000 - \left(\frac{(1.0075)^{36} - 1}{0.0075}\right) \times M = 0$$

Therefore, we can solve for M:

$$M = \left(\frac{0.0075}{(1.0075)^{36} - 1}\right) \times (1.0075)^{36} \times 2{,}000$$

Note: It might be useful in working with the preceding equations to review multiplication of fractions.

In any case, plugging into our calculator, we quickly find that the required payment will be $63.60 per month. Hence the total payment will be:

$63.60 \times 36 = \$2{,}289.60$

This amount is much less than if you were to pay compound interest on the entire $2,000 for the entire 36 months, as shown in the following equation:

$$(1.0075)^{36} \times \$2{,}000 = \$2{,}617.29$$

Appendix D

Resource for Secondary School Teachers: Circumcenter, Orthocenter, and Centroid

The purpose of this appendix is to give a demonstration—albeit on a small scale—of how the usual tedium and pitfalls of the axiomatic development of Euclidean geometry might be avoided. It deals with very standard materials: why the perpendicular bisectors (resp., *altitudes* and *medians*) of a triangle must meet at a point, the *circumcenter* (resp., *orthocenter* and *centroid*) of the triangle. The exposition starts at the lowest level—the axioms—and ends with a proof of the concurrence (i.e., meeting at a point) of the medians. It also includes a collection of exercises on proofs as an indication of how and when such exercises might be given. The major theorems to be proved (Theorems 1 and 11–14) are all interesting and are likely regarded as surprising to most students. These theorems would therefore do well to hold students' attention and convince students of the value of mathematical proofs.

The goal of this appendix is to prove the concurrence of the medians. If one turns this proof "inside out," so to speak, one will get the proof of the concurrence of the altitudes. The proof of the latter theorem is also included as is a demonstration of the concurrence of the perpendicular bisectors since that is also needed. The fact that the concurrence of the angle bisectors (the *incenter*) is left out is therefore entirely accidental. This appendix makes no pretense at completeness because its only purpose is to demonstrate a particular approach to geometric exposition, but if it did, then certainly the four centers would have been discussed together.

We specifically call attention to the following features:

1. The appearance of the exercises on proofs is intentionally gauged to approximate at what point of the axiomatic development those exercises should be given to students in a classroom situation. The first of such exercises asks only for a straightforward imitation of a proof that has just been presented (Lemma (2B)). The next one asks only for the reasons for some steps in the proof of Lemma 6, and by then students have already been exposed to several nontrivial proofs. The first exercise that asks for a genuine proof occurs after Lemma 7. In other words, students are given ample time to absorb the idea of a proof by studying several good examples before they are asked to construct one by themselves.

   A conscientious effort was also made to ensure that the exercises all have some geometric content so that any success with them would require some geometric understanding instead of just facility with formal reasoning.

Date: *October 18, 1998.*

2. Certain facts are explicitly assumed without proof before some of the proofs (*local axiomatics*). Students should be informed that they too can make use of these unproven assertions.
3. None of the results presented is trivial to a beginner (except Lemma 4). It is hoped that, altogether, these results will convince the students of the benefit of learning about proofs; namely, to understand why some interesting things are true and be able, in turn, to present arguments to convince other people. In fact, most students probably do not believe any of the major theorems (Theorems 1 and 11–14) before being exposed to their proofs.
4. The concurrence of the altitudes and medians (Theorems 11 and 14) is usually not presented in standard textbooks except by use of coordinate geometry or the concept of similarity. Thus those theorems tend not to appear in a typical geometry curriculum. However, the easy access to them as demonstrated by this appendix should be a convincing argument that, with a little effort, it is possible to present students with interesting results very early in a geometry course.
5. The two-column proofs given in the following pages most likely do not conform to the rigid requirements imposed on the students in some classrooms (cf. Schoenfeld 1988, 145–66). However, for exposition, they are perfectly acceptable by any mathematical standards. It is hoped that their informal character would help restore the main focus of a proof, which is the correctness of the mathematical reasoning instead of a rigidly correct exposition.
6. The proof of Theorem 12 is given twice: once in the two-column format and the second time in the narrative (paragraph) format. In the classroom such "double-proofs" should probably be done for a week or two to lead students away from the two-column format. The proof of Theorem 13 is given only in the narrative format.

Axioms

We shall essentially assume the School Mathematics Study Group (SMSG) axioms, which are *paraphrased* below rather than quoted verbatim for easy reference; the relevant definitions are usually omitted (see Cederberg 1989, 210–11). Only those axioms pertaining to *plane* Euclidean geometry are given. Moreover, a school geometry course has no time for a *minimum* set of axioms. The last three axioms have therefore been added to speed up the logical development.

1. Two points A and B determine a unique line, to be denoted by AB.
2. (The Distance Axiom). To every pair of distinct points there corresponds a unique positive number, called their *distance*. This distance satisfies the requirement of the next axiom.
3. (The Ruler Axiom). Every line can be put in one-one correspondence with the real numbers so that if P and Q are two points on the line, then the absolute value of the difference of the corresponding real numbers is the distance between them.

4. (The Ruler Placement Axiom). Given two points P and Q on a line, the correspondence with real numbers in the preceding axiom can be chosen so that P corresponds to zero and Q corresponds to a positive number.
5. There are at least three noncollinear points.
6. (The Plane Separation Axiom). Given a line ℓ. Then the points not on ℓ form two convex sets, and any line segment $\overline{AB}$ joining a point A in one set and a point B in the other must intersect ℓ. The convex sets are called the *half-planes* determined by ℓ.
7. (The Angle Measurement Axiom). To every $\angle ABC$ there corresponds a real number between 0 and 180, to be denoted by $m\angle ABC$, called the *measure* of the angle.
8. (The Angle Construction Axiom). Given a line AB and a half-plane H determined by AB, then for every number r between 0 and 180, there is exactly one ray $\overrightarrow{AP}$ in H so that $m\angle PAB = r$.
9. (The Angle Addition Axiom). If D is a point in the interior of $\angle BAC$, then $m\angle BAC = m\angle BAD + m\angle DAC$.
10. (The Angle Supplement Axiom). If two angles form a linear pair, then their measures add up to 180.
11. SAS Axiom for congruence of triangles.
12. (The Parallel Axiom). Through a given external point, there is at most one line parallel to a given line.
13. (The Area Axiom). To every polygonal region, there corresponds a unique positive number, called its *area*, with the following properties: (i) congruent triangles have the same area; (ii) area is additive on disjoint unions; and (iii) the area of a rectangle is the product of the lengths of its sides.
14. SSS Axiom for congruence of triangles.
15. ASA Axiom for congruence of triangles.
16. (The AA Axiom for Similarity). Two triangles with two pairs of angles equal are similar.

Perpendicularity

In the following exposition, we shall abbreviate "$m\angle ABC$" to simply "$\angle ABC$." So if $\angle ABC$ has measure 45, we write $\angle ABC = 45$.

Recall that $\angle DCB$ is a *right angle*, (see figure 1) and DC is *perpendicular* to CB, if for a point A collinear with C and B and on the other side of C from B, $\angle DCA = \angle DCB$. If $\angle DCB$ is a right angle, then its measure is 90 because by the Angle Supplement Axiom $\angle DCA + \angle DCB = 180$ so that $\angle DCB + \angle DCB = 180$, and we obtain $\angle DCB = 90$.

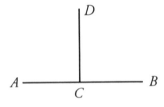

Figure 1

Similarly, $\angle DCA = 90$. Conversely, if A, C, B are collinear and $\angle DCB = 90$, the same argument shows $\angle DCA = 180 - \angle DCB = 90$, and $\angle DCB = \angle DCA$ so that $\angle DCB$ is a right angle. Thus we can assert that two lines ℓ_1, ℓ_2 are perpendicular if one of the angles they form is 90.

Recall also that a point C on a segment $\overline{AB}$ is the *midpoint* of $\overline{AB}$ if $\overline{AC} = \overline{CB}$. (Recall the *convention*: AB denotes the line containing A and B, and $\overline{AB}$ denotes the line *segment* joining A to B.) The straight line passing through the midpoint of a segment and perpendicular to it is called the *perpendicular bisector* of the segment. Note that every segment has a perpendicular bisector. Indeed, given $\overline{AB}$, the Ruler Axiom guarantees that there is a midpoint C of $\overline{AB}$, and the Angle Construction Axiom guarantees that there is an $\angle DCB = 90$. Then by the preceding discussion, $DC \perp AB$, and DC is the perpendicular bisector.

We shall need the fact that the perpendicular bisector of a segment is unique; i.e., if DC and $D'C'$ are perpendicular bisectors of $\overline{AB}$, then $DC = D'C'$. This is so intuitively obvious that we shall not spend time to prove it.

> [For those interested in a proof, however, one uses the Distance Axiom and the Ruler Axiom to show that the midpoint of a segment must be unique, and then one uses the Angle Construction Axiom to show that the line passing through the midpoint and perpendicular to the segment is also unique.]

Three lines are *concurrent* if they meet at a point. The following gives a surprising property about perpendicular bisectors:

THEOREM 1. *The perpendicular bisectors of the three sides of a triangle are concurrent.* (The point of concurrency is called the *circumcenter* of the triangle.)

Let A', B', C' be the midpoints of $\overline{BC}$, $\overline{AC}$, and $\overline{AB}$, resp. (see figure 2). A naive approach would try to prove directly that all three perpendicular bisectors meet at a point O. This is too clumsy and also unnecessary. A better way is the following. Take two of the perpendicular bisectors, say, those at A' and B', and let them meet at a point O. Then we show that O must lie on the perpendicular bisector of $\overline{AB}$. Theorem 1 would be proved.

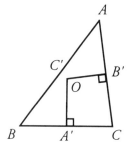

Figure 2

One would appreciate this approach to the proof of Theorem 1 more if the perpendicular bisector of a segment is better understood. To this end we first prove:

LEMMA 2. *A point D is on the perpendicular bisector of a segment $\overline{AB}$ if and only if $\overline{DA} = \overline{DB}$.*

PROOF. First, we explain the "if and only if" terminology. It is a shorthand to indicate that two assertions must be proved:

(i) *If the statement preceding this phrase is true, then the statement following this phrase is also true.*

(ii) *If the statement following this phrase is true, then the statement preceding this phrase is also true.*

For the case at hand, this means we have to prove two things (see figure 3):

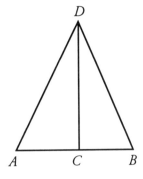

Figure 3

LEMMA (2A). *If D is on the perpendicular bisector of $\overline{AB}$, then $\overline{DA} = \overline{DB}$.*
LEMMA (2B). *If $\overline{DA} = \overline{DB}$, then D is on the perpendicular bisector of $\overline{AB}$.*

PROOF OF (2A): Let CD be the perpendicular bisector of $\overline{AB}$.

1. $\overline{AC} = \overline{CB}$, and $\angle DCA = \angle DCB = 90$. 1. Hypothesis.
2. $\overline{CD} = \overline{CD}$. 2. Obvious.
3. $\triangle ADC \cong \triangle BDC$. 3. SAS.
4. $\overline{DA} = \overline{DB}$. 4. Corresponding sides of congruent triangles. Q.E.D.

PROOF OF (2B): Given that $\overline{DA} = \overline{DB}$, we have to show that the perpendicular bisector of $\overline{AB}$ passes through D. Instead of doing so directly, we do something rather clever: we are going to construct the *angle bisector CD* of $\angle ADB$. This means $\angle ADC = \angle CDB$. Of course, we must first prove that there is such a line CD with the requisite property. Then we shall prove $CD \perp AB$ and $\overline{AC} = \overline{CB}$ so that CD is the perpendicular bisector of $\overline{AB}$.

Recall that the Plane Separation Axiom makes it possible to define the *interior* of $\angle ADB$ as the intersection of the half-plane determined by DA which contains B and the half-plane determined by DB which contains A. Now the following assertion is obvious pictorially:

ASSERTION A. *If E is a point such that $\angle ADE < \angle ADB$, then E is in the interior of $\angle ADB$. Furthermore, the line DE intersects $\overline{AB}$.*

For those who are truly curious, let it be mentioned that the first part of Assertion A can be proved by using the Angle Addition Axiom and the second part by repeated applications of the Plane Separation Axiom (see figure 4).

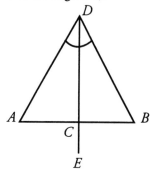

Figure 4

1. If the measure of $\angle ADB$ is x, there is a point E in the interior of $\angle ADB$ so that $\angle ADE = \frac{x}{2}$.
2. If DE meets $\overline{AB}$ at C, $\angle ADC + \angle CDB = \angle ADB = x$.
3. $\angle CDB = x - \angle ADC = x - \frac{x}{2} = \frac{x}{2}$.
4. $\angle ADC = \angle CDB$.
5. $\overline{DA} = \overline{DB}$.
6. $\overline{CD} = \overline{CD}$.
7. $\triangle ACD \cong \triangle BCD$.
8. $\overline{AC} = \overline{CB}$ and $\angle DCA = \angle DCB$.
9. $\overline{CD}$ is the perpendicular bisector of $\overline{AB}$.

1. Angle Construction Axiom and Assertion A.

2. Angle Addition Axiom and Assertion A.

3. By 1 and 2.
4. By 1 and 3.
5. Hypothesis.
6. Obvious.
7. SAS.
8. Corresponding angles and sides of congruent triangles.

9. By 8 and the definition of a perpendicular bisector. Q.E.D.

Exercise 1. Using the preceding proof as a model, write out a complete proof of the fact that if $\triangle ABC$ is isosceles with $\overline{AB} = \overline{AC}$, then the angle bisector of $\angle A$ is also the perpendicular bisector of $\overline{BC}$.

From the point of view of Lemma 2, our approach to the proof of Theorem 1 is now more transparent. As mentioned previously, this approach is to let the perpendicular bisectors $A'O$ and $B'O$ of $\overline{BC}$ and $\overline{AC}$, resp., meet at a point O. Then we shall prove that O lies on the perpendicular bisector of $\overline{AB}$ (see figure 5).

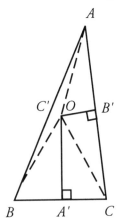

Figure 5

PROOF OF THEOREM 1. Join OB, OC, and OA.

1. $\overline{OB} = \overline{OC}$.　　　　　　　1. Lemma (2A) and the fact that $\overline{OA'}$ is the perpendicular bisector of $\overline{BC}$.
2. $\overline{OC} = \overline{OA}$.　　　　　　　2. Lemma (2A) and the fact that $\overline{OB'}$ is the perpendicular bisector of $\overline{AC}$.
3. $\overline{OB} = \overline{OA}$.　　　　　　　3. From 1 and 2.
4. O lies on the perpendicular bisector of $\overline{AB}$.　　　　4. Lemma (2B). Q.E.D.

Circumcenter of a Triangle

COROLLARY TO THEOREM 1. *The circumcenter of a triangle is equidistant from all three vertices.*

The corollary is obvious if we look at steps 1–3 of the preceding proof. Note that, as figure 6 suggests, the circumcenter can be in the exterior of the triangle.

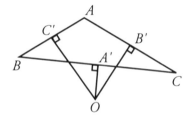

Figure 6

Several remarks about the proof of Theorem 1 are in order. First of all, this proof suggests a general method of proving the concurrency of three lines: let two of the three meet at a point O, and then show that O must lie on the third line. Technically, this is easier than directly proving that all three lines meet at a point. This method is a kind of *indirect proof* and is useful in many situations; for example, in proving that the three angle bisectors of a triangle are concurrent. In other words, if we think of proving a theorem as fighting a battle against an enemy, then it makes sense that *sometimes* we can defeat the enemy without resorting to a frontal attack.

A second remark has to do with the tacit assumption above, namely, that the perpendicular bisectors $\overline{OA'}$ and $\overline{OB'}$ of $\overline{BC}$ and $\overline{AC}$, resp., do meet at a point O. This is obvious from figure 6, and we usually do not bother to prove such obvious statements, being fully confident that—if challenged—we can prove them. For the sake of demonstration, however, we will supply a proof this time after we have proved a few properties of parallel lines. Thus, we shall prove:

ASSERTION B. *Perpendicular bisectors from two sides of a triangle must intersect.*

Note that there is no circular reasoning here: Assertion B will not be used to prove any of the theorems involving parallel lines. Indeed, we shall not have to face Assertion B again in this appendix.

A third remark concerns the name *circumcenter*. A *circle* with center O and radius r is by definition the collection of all points whose distance from O is r. The corollary to Theorem 1 may then be rephrased as: the circle with center O and radius $\overline{OA}$ passes through all three vertices. This circle is called the *circumcircle* of $\triangle ABC$, which then gives rise to the name "circumcenter." (*Circum* means "around.") Incidentally, Theorem 1 proves that any triangle determines a circle that passes through all three vertices.

Next, we turn attention to the *altitudes* of a triangle; i.e., the perpendiculars from the vertices to the opposite sides (see figure 7). We want to show that they too are concurrent. This demonstration needs some preparation. First of all, we have to show that altitudes exist; i.e., through each vertex there is a line that is perpendicular to the opposite side. More generally, we shall prove:

LEMMA 3. *Given a point P and a line ℓ not containing P, there is a line PQ which is perpendicular to ℓ.*

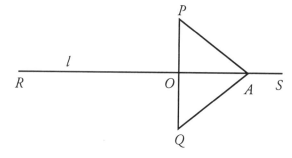

Figure 7

PROOF. Let $\ell = RS$. Recall that ℓ contains an infinite number of points (the Ruler Axiom), and that the Plane Separation Axiom allows us to talk about the two sides of ℓ.

1. Join P to an arbitrary point A on ℓ.   1. Two points determine a line.
2. If x is the measure of $\angle PAR$, let Q be a point on the side of ℓ not containing P so that $\angle PAQ = x$.   2. The Angle Construction Axiom.
3. We may let Q be the point on AQ so that Q, P are on opposite sides of ℓ and $\overline{AQ} = \overline{AP}$.   3. The Distance Axiom.
4. PQ meets ℓ at some point O.   4. The Plane Separation Axiom.
5. $\angle PAO = \angle QAO$.   5. From 1 and 2.
6. $\overline{OA} = \overline{OA}$.   6. Obvious.
7. $\triangle PAO \cong \triangle QAO$.   7. SAS.
8. $\angle AOP = \angle AOQ$.   8. Corresponding angles of congruent triangles.
9. $PQ \perp \ell$.   9. By definition of perpendicularity. Q.E.D.

Vertical Angles

Before we turn to parallel lines, we do some spadework. The teacher introduces the definition of *vertical angles* (omitted here).

LEMMA 4. *Vertical angles are equal.*

PROOF. Let AB, CD meet at O. We will show $\angle AOD = \angle BOC$ (see figure 8).

1. $\angle AOD + \angle DOB = 180$ and $\angle DOB + \angle BOC = 180$.   1. The Angle Supplement Axiom.
2. $\angle AOD + \angle DOB = \angle DOB + \angle BOC$.   2. By 1.
3. $\angle AOD = \angle BOC$.   3. From 2 and the cancellation law of addition. Q.E.D.

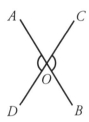

Figure 8

The teacher introduces the definitions of *exterior angle* and *remote interior angles* of a triangle (omitted here). To prove the next proposition, we shall assume a geometrically obvious fact. In figure 9 if M is any point on $\overline{AC}$, we shall assume as known:

ASSERTION C. *If we extend $\overline{BM}$ along M to a point E, then E is always in the interior of $\angle ACD$.*

This can be proved with repeated applications of the Plane Separation Axiom, but the argument is not inspiring.

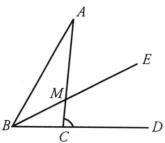

Figure 9

PROPOSITION 5. *An exterior angle of a triangle is greater than either remote interior angle.*

PROOF. Let us show $\angle ACD > \angle BAC$ (see figure 10). To show $\angle ACD > \angle ABC$, we observe that the same proof would show $\angle BCG > \angle ABC$ and then use Lemma 4 to get $\angle BCG = \angle ACD$. Putting the two facts together, we get $\angle ACD > \angle ABC$.

Join B to the midpoint M of $\overline{AC}$ and extend BM to a point E such that $\overline{BM} = \overline{ME}$ (the Ruler Axiom). Join CE.

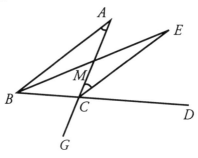

Figure 10

1. $\angle AMB = \angle EMC$.
2. $\overline{BM} = \overline{ME}$.
3. $\overline{AM} = \overline{MC}$.
4. $\triangle AMB \cong \triangle CME$.
5. $\angle BAM = \angle MCE$.

6. $\angle MCE < \angle ACD$.

7. $\angle BAC < \angle ACD$.

1. Lemma 4.
2. By construction.
3. M is the midpoint of $\overline{AC}$.
4. SAS.
5. Corresponding angles of congruent triangles.
6. By the Angle Addition Axiom and Assertion C.
7. By 5 and 6. Q.E.D.

Parallel Lines

We now come to some basic facts about parallel lines. Given two lines ℓ_1 and ℓ_2, one can introduce the definition of *alternate interior angles* and *corresponding angles* of ℓ_1 and ℓ_2 with respect to a transversal (omitted here). We shall need:

LEMMA 6. *If two lines make equal alternate interior angles with a transversal, they are parallel.*

PROOF. Let the transversal be BE. Designate the two equal alternate interior angles as $\angle\alpha$ and $\angle\beta$ (see figure 11). We assume that AC is not $\parallel$ to DF and deduce a contradiction. (This is an example of *proof by contradiction*.)

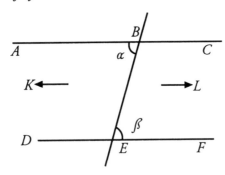

Figure 11

1. AC meets DF either at a point K to the left of BE or at L to the right of BE.
2. If AC meets DF at K, then $\angle\beta > \angle\alpha$, contradicting $\angle\alpha = \angle\beta$.
3. If AC meets DF at L, then $\angle\alpha > \angle\beta$, also contradicting $\angle\alpha = \angle\beta$.

1. By the fact that AC is not $\parallel$ to DF and by the Plane Separation Axiom.
2. ☐
3. ☐ Q.E.D.

Exercise 2. Supply the reasons for steps 2 and 3.
[Answers: Proposition 5 is the reason for both.]

Note: The textbook and teacher must make sure that students are eventually given the answers to problems of this nature; it is important to bring closure to a mathematical discussion.

This proposition complements the parallel axiom in the following sense. Notation is given as in the preceding proof: suppose that DF and B are given and we want to construct a line through B and $\parallel DF$. By the Angle Construction Axiom, with $\angle\beta$ as given, we can construct $\angle\alpha$ with vertex at the given B on the other side of $\angle\beta$ but with the same measure. Then by Lemma 6, AC is a line passing through B which is $\parallel DF$. Therefore:

COROLLARY TO LEMMA 6. *Through a point not on a line ℓ, there is one and only one line parallel to ℓ.*

Lemma 7 is the converse of Lemma 6.

LEMMA 7. *Alternate interior angles of parallel lines with respect to a transversal are equal.*

PROOF. The notation is as before, suppose $AC \parallel DF$. We shall prove $\angle \alpha = \angle \beta$ (see figure 12).

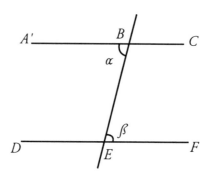

Figure 12

1. At B, construct $\angle A'BE$ to the left of BE so that $\angle A'BE = \angle \beta$.
2. $A'B \parallel DF$.
3. Since $A'B$ passes through B, $A'B = AB$.
4. $\angle \alpha = \angle A'BE = \angle \beta$.

1. Angle Construction Axiom.
2. ☐
3. ☐
4. By 3. Q.E.D.

Exercise 3. Supply the reasons for steps 2 and 3.
[Answers: Step 2. Lemma 6.   Step 3. The Parallel Axiom.]

Exercise 4. Prove that corresponding angles of parallel lines with respect to a transversal are equal.

Parallelograms

The teacher introduces the definition of a *quadrilateral* (omitted here). A *parallelogram* is a quadrilateral with parallel opposite sides. We shall need two properties of parallelograms that are pictorially plausible when a parallelogram is drawn carefully.

LEMMA 8. *A quadrilateral is a parallelogram if and only if it has a pair of sides which are parallel and equal.*

LEMMA 9. *A quadrilateral is a parallelogram if and only if its opposite sides are equal.*

PROOF OF LEMMA 8. First we prove that if quadrilateral $ABCD$ has a pair of sides which are parallel and equal, then it is a parallelogram. In figure 13 we assume $\overline{AB} = \overline{CD}$ and $\overline{AB} \parallel \overline{CD}$. Then we have to prove $\overline{AD} \parallel \overline{BC}$.

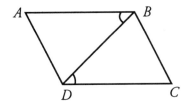

Figure 13

1. Join BD. $\overline{BD} = \overline{BD}$.	1. Two points determine a line.
2. $\angle ABD = \angle BDC$.	2. Lemma 7.
3. $\triangle ABD \cong \triangle CDB$.	3. SAS.
4. $\angle ADB = \angle DBC$.	4. Corresponding angles of congruent triangles.
5. $AD \parallel BC$.	5. Lemma 6. Q.E.D.

Next we prove that a parallelogram has a pair of sides which are parallel and equal. Since $AB \parallel DC$ by definition, it suffices to prove that $\overline{AB} = \overline{DC}$. Let notation be as in the preceding proof.

1. $\angle ABD = \angle BDC$ and $\angle ADB = \angle DBC$.	1. Lemma 7.
2. $\overline{BD} = \overline{BD}$.	2. Obvious.
3. $\triangle ABD \cong \triangle CDB$.	3. ASA.
4. $\overline{AB} = \overline{DC}$.	4. Corresponding sides of congruent triangles. Q.E.D.

Exercise 5. Prove Lemma 9 (it is similar to the proof of Lemma 8).

Exercise 6. Prove that a quadrilateral $ABCD$ is a parallelogram if and only if the diagonals $\overline{AC}$ and $\overline{BD}$ bisect each other; i.e., if they intersect at E, then $\overline{AE} = \overline{EC}$ and $\overline{BE} = \overline{ED}$ (see figure 14).

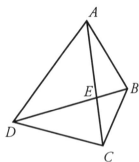

Figure 14

The following lemma is an immediate consequence of Lemma 6. Convince yourself of this, and be sure to draw pictures to see what it says.

LEMMA 10. *Suppose two lines ℓ_1 and ℓ_2 are parallel. (i) If ℓ is a line perpendicular to ℓ_1, then ℓ is also perpendicular to ℓ_2. (ii) If another two lines L_1 and L_2 satisfy $L_1 \perp \ell_1$ and $L_2 \perp \ell_2$, then $L_1 \parallel L_2$.*

We can now prove Assertion B stated after the proof of Theorem 1. Let $\triangle ABC$ be given and let lines ℓ_1 and ℓ_2 be the perpendicular bisectors of $\overline{BC}$ and $\overline{AC}$, resp. (see figure 15). Let lines L_1 and L_2 be lines containing $\overline{BC}$ and $\overline{AC}$, resp. If Assertion B is false, then $\ell_1 \parallel \ell_2$. By Lemma 10 (ii), $L_1 \parallel L_2$. But we know L_1 meets L_2 at C, a contradiction. Thus ℓ_1 must meet ℓ_2 after all.

We are in a position to prove one of our main results.

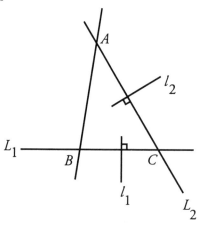

Figure 15

The Orthocenter

THEOREM 11. *The three altitudes of a triangle are concurrent.* (This point is called the orthocenter of the triangle.)

PROOF. Let $\triangle ABC$ be given and let its altitudes be AD, BE, and CF. The idea of the proof is to turn AD, BE, CF into perpendicular bisectors of a bigger triangle and use Theorem 1. The idea itself is sophisticated and is attributed to the great mathematician C. F. Gauss. Technically, however, it is quite simple to execute. It illustrates a general phenomenon in mathematics: sometimes a seemingly difficult problem becomes simple when it is put into the proper context (see figure 16).

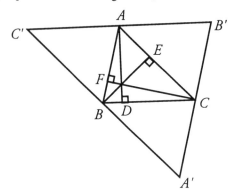

Figure 16

1. Through A, there is a line $\overline{C'B'} \parallel \overline{BC}$.	1. Corollary to Lemma 6.
2. Similarly, let $C'A'$ and $B'A'$ be lines through B and C, resp., such that $\overline{C'A'} \parallel \overline{AC}$ and $\overline{B'A'} \parallel \overline{AB}$.	2. Same reason.
3. $ABCB'$ is a parallelogram.	3. From 1 and 2.
4. $\overline{AB'} = \overline{BC}$.	4. Lemma 9.
5. $ACBC'$ is likewise a parallelogram and $\overline{C'A} = \overline{BC}$.	5. See 3 and 4.
6. $\overline{C'A} = \overline{AB'}$.	6. From 4 and 5.
7. $AD \perp BC$.	7. Hypothesis.
8. $AD \perp C'B'$.	8. Lemma 10(i).
9. AD is the perpendicular bisector of $\overline{C'B'}$.	9. From 6 and 8.
10. Similarly, BE and CF are perpendicular bisectors of $\overline{C'A'}$ and $\overline{A'B'}$, resp.	10. See 3 through 9.
11. AD, BE, and CF are concurrent.	11. Apply Theorem 1 to $\triangle A'B'C'$. Q.E.D.

The Medians and Centroid of a Triangle

The line joining a vertex of a triangle to the midpoint of the opposite side is called a *median* of the triangle.

Exercise 7. In the notation of the proof of Theorem 11, prove that AA', BB', and CC' are the medians of $\triangle ABC$ as well.

Finally, we turn to the proof of the concurrence of the medians. The proof will be seen to have many points of contact with the proof of Theorem 11 shown above. Instead of turning "outward" to a bigger triangle, however, the proof of the concurrence of the medians turns "inward" and looks at the triangle obtained by joining the midpoints of the three sides. To this end, the following theorem is fundamental.

THEOREM 12. *The line segment joining the midpoints of two sides of a triangle is parallel to the third side and is equal to half of the third side.*

PROOF. Thus if $\overline{AE} = \overline{EB}$ and $\overline{AF} = \overline{FC}$, then $\overline{EF} \parallel \overline{BC}$ and $\overline{EF} = \tfrac{1}{2}\overline{BC}$.

To motivate the proof, note that all the axioms and the theorems presented so far deal with the equality of two objects (angles, segments, and so forth), not about *half* of something else. So it makes sense to try to reformulate $\overline{EF} = \tfrac{1}{2}\overline{BC}$ as a statement about the equality of two equal segments (see figure 17). What then is simpler than doubling $\overline{EF}$? Students will learn that the construction of so-called auxiliary lines, such as $\overline{FP}$ and $\overline{PC}$ in the following proof, is a fact of life in Euclidean geometry.

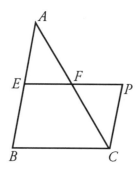

Figure 17

1. Extend $\overline{EF}$ along F to P so that $\overline{EF} = \overline{FP}$ and join PC.
2. $\overline{AF} = \overline{FC}$.
3. $\angle AFE = \angle PFC$.
4. $\triangle AFE \cong \triangle CFP$.
5. $\angle AEF = \angle FPC$.

6. $EB \parallel PC$.
7. $\overline{AE} = \overline{PC}$.

8. $\overline{AE} = \overline{EB}$.
9. $\overline{EB} = \overline{PC}$.
10. $EBCP$ is a parallelogram.
11. $EF \parallel BC$.

12. $\overline{EP} = \overline{BC}$.
13. $\overline{EP} = 2\,\overline{EF}$.
14. $\overline{EF} = \tfrac{1}{2}\,\overline{BC}$.

1. The Ruler Axiom and the fact that two points determine a line.
2. Hypothesis.
3. Lemma 4.
4. ☐
5. By 4 and corresponding angles of congruent triangles are equal.
6. Lemma 6.
7. By 4 and corresponding sides of congruent triangles are equal.
8. Hypothesis.
9. By 7 and 8.
10. ☐
11. By 10 and the definition of a parallelogram.
12. ☐
13. By 1.
14. By 12 and 13. Q.E.D.

Exercise 8. Supply the reasons for steps 4, 10, and 12.
[Answers: Step 4. By 1 through 3 and SAS. Step 10. By 6, 9, and Lemma 8. Step 12. By 10 and Lemma 9.]

PROOF OF THEOREM 12 IN NARRATIVE FORM. Extend $\overline{EF}$ to a point P so that $\overline{EF} = \overline{FP}$. Join PC. We are going to prove that $\triangle AEF \cong \triangle CPF$. This proof is possible because the vertical angles $\angle AFE$ and $\angle CFP$ are equal, $\overline{AF} = \overline{FC}$ by hypothesis and $\overline{EF} = \overline{FP}$ by construction. So SAS gives the desired congruence. It follows that $\angle AEF = \angle CPF$ and therefore that $AB \parallel PC$ (Lemma 6) and $\overline{AE} = \overline{PC}$. Because $\overline{AE} = \overline{EB}$, EB and PC are both parallel and equal. Hence $EBCP$ is a parallelogram (Lemma 8).

In particular, $EP \parallel BC$ and, by Lemma 10, $\overline{EP} = \overline{BC}$. Hence $\overline{BC} = \overline{EF} + \overline{FP} = 2\overline{EF}$. Q.E.D.

Exercise 9. Prove that two lines that are each parallel to a third line are parallel to each other.

In the next four exercises, do not use Axiom 16 on similarity (p. D-3) *in your proofs.*

Exercise 10. Let E be the midpoint of $\overline{AB}$ in $\triangle ABC$. Then the line passing through E which is parallel to BC bisects AC.

Exercise 11. Let $ABCD$ be *any* quadrilateral, and let A', B', C', and D' be the midpoints of $\overline{AB}$, $\overline{BC}$, $\overline{CD}$, and $\overline{DA}$, resp. Then $A'B'C'D'$ is a parallelogram.

Exercise 12. Given $\triangle ABC$. Let L, M be points on $\overline{AB}$ and $\overline{AC}$, resp., so that $\overline{AL} = \frac{1}{4}\overline{AB}$ and $\frac{1}{4}\overline{AB}$ and $\overline{AM} = \frac{1}{4}\overline{AC}$. Prove that $LM \parallel BC$ and $\overline{LM} = \frac{1}{4}\overline{BC}$.

Exercise 13. Given $\triangle ABC$. Let L, M be points on $\overline{AB}$ and $\overline{AC}$, resp., so that $\overline{AL} = \frac{1}{3}\overline{AB}$ and $\overline{AM} = \frac{1}{3}\overline{AC}$. Prove that $LM \parallel BC$ and $\overline{LM} = \frac{1}{3}\overline{BC}$. (Hint: Begin by imitating the proof of Theorem 12.)

Exercise 14. (For those who know mathematical induction.) Let n be a positive integer. Given $\triangle ABC$. Let L, M be points on $\overline{AB}$ and $\overline{AC}$, resp., so that $\overline{AL} = \frac{1}{n}\overline{AB}$ and $\overline{AM} = \frac{1}{n}\overline{AC}$. Prove that $LM \parallel BC$ and $\overline{LM} = \frac{1}{n}\overline{BC}$.

For the proof of the next theorem, we shall assume the following three facts:

> **ASSERTION D.** *Any two medians of a triangle meet in the interior of the triangle.*
> **ASSERTION E.** *Two lines that are each parallel to a third are parallel to each other.*
> **ASSERTION F.** *The diagonals of a parallelogram bisect each other.*

For Assertion D one first has to define the *interior* of a triangle by using the Plane Separation Axiom. The proof then uses this axiom repeatedly, a tedious process. Assertion E is Exercise 9 shown above, and Assertion F is Exercise 6 on p. D-13.

THEOREM 13. *Let BE be a median of $\triangle ABC$. Then any other median must meet $\overline{BE}$ at the point G so that $\overline{BG} = 2\overline{GE}$.*

PROOF. Let CF be another median and let CF meet BE at a point to be denoted also by G for simplicity. We will prove that $\overline{BG} = 2\overline{GE}$, which would then finish the proof of the theorem (see figure 18).

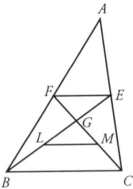

Figure 18

Join FE and join the midpoint L of $\overline{BG}$ to the midpoint M of $\overline{CG}$. Applying Theorem 12 to $\triangle ABC$, we get $\overline{BC} = 2\,\overline{FE}$ and $FE \parallel BC$. Similiarly, applying the same theorem to $\triangle GBC$ yields $\overline{BC} = 2\,\overline{LM}$ and $LM \parallel BC$. Hence $\overline{FE}$ and $\overline{LM}$ are equal and parallel (by Assertion E), and $FEML$ is a parallelogram (Lemma 8). By Assertion F, $\overline{LG} = \overline{GE}$. L being the midpoint of $\overline{BG}$ implies $\overline{BL} = \overline{LG} = \overline{GE}$ so that $\overline{BG} = 2\,\overline{GE}$. Q.E.D.

Exercise 15. Let D, E, F be the midpoints of $\overline{BC}, \overline{AC},$ and $\overline{AB}$, resp., in $\triangle ABC$. Prove that $\triangle AFE, \triangle DFE, \triangle FBD,$ and $\triangle EDC$ are all congruent.

$\triangle DFE$ in Exercise 15 is called the *medial triangle* of $\triangle ABC$.

It follows immediately from Theorem 13 that both of the other two medians of $\triangle ABC$ must intersect $\overline{BE}$ at the point G. Hence we have:

THEOREM 14. *The three medians of a triangle are concurrent, and the point of concurrency is two-thirds of the length of each median from the vertex.* (This is the *centroid* of the triangle.)

Exercise 16. Show that the centroid of a triangle is also the centroid of its medial triangle.

One may conjecture in view of Theorem 14 that if we trisect each side of a triangle, then the lines joining a vertex to an appropriate point of trisection on the opposite side may also be concurrent. One accurately drawn picture (see figure 19) is enough to lay such wishful thinking to rest. Such a picture then provides a *counterexample* to this conjecture.

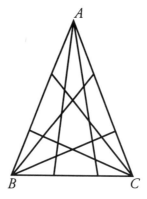

Figure 19

Part of the charm of Euclidean geometry is that most conjectures can be made plausible or refuted by a judicious picture. Compared with other subjects, such as algebra or calculus, this way of confronting a conjecture in geometry is by far the most pleasant.

Appendix E
Sample Mathematics Problems

The sample problems in this appendix supplement those appearing in Chapter 2. The problems for kindergarten through grade seven are organized by grade level and according to the standards for the five strands in the *Mathematics Content Standards*. The problems for grades eight through twelve are presented according to the standards for the discipline headings algebra, geometry, and so forth. Teachers may use these problems as a resource to develop students' skills in working with the standards. [Complete citations for the sources following some of the mathematics problems in this appendix appear in "Works Cited." Many of the problems come from or are adapted from materials that are a part of the Third International Study of Mathematics and Science (TIMSS). TIMSS offers both a resource kit, *Attaining Excellence: A TIMSS Resource Kit,* and a Web site <http://www.csteep.bc.edu/TIMSS1/pubs_main.html>. *Ed.*]

Kindergarten

Number Sense

1.1, 1.2 How many students are in your class? How many chairs are in the room? Are there more chairs than students? What happens when there are more students than chairs?

1.2 Make sure that students can count forward accurately before presenting a series like the following:

30, 29, 28, __, __, 25, __, __, __, __, 20, 19, __, __, __, 15.

Grade One

Number Sense

1.2 Prove or disprove a classmate's claim that, "29 is more than 41 because 9 is more than 4 or 1." (This problem also applies to Mathematical Reasoning Standard 2.1.)

Grade Two

Number Sense

2.2 Find a three-digit number such that the sum of its digits is equal to 26. How many such numbers can you find?

3.2   Pretend you are at a store and you have $2.00 to spend. A pen costs 79 cents, a notepad 89 cents, and an eraser 49 cents. Suppose you want to buy two items and have the most money left over, which two would you buy? What is the largest number of pens you can buy? Notepads? Erasers? Explain how you know.

5.2   How many pennies does it take to make $1.57? How many nickels does it take to make $2.65?

Measurement and Geometry

1.1   Which is longer: the width of your classroom or 8 times the length of your desk?

Statistics, Data Analysis, and Probability

1.1   A kite has four panels. You have been asked to color it with either red or blue on each panel. How many different color kites can you make?

2.1   Look at these numbers: 50, 46, 42, 38, 34, 30, . . . There are many patterns that can produce these numbers. Please describe one. (Teacher: follow up with a question about which method is the simplest.) (Adapted from TIMSS gr. 4, U-4)

Grade Three

Number Sense

2.4   There are 54 marbles. They are put into 6 bags so that the same number of marbles is in each. How many marbles would 2 bags contain? (Adapted from TIMSS gr. 4, K-9)

2.8   Here is a number sentence: 2,000 + ___ + 30 + 9 = 2,739. What number should go where the blank is to make the sentence true? (TIMSS gr. 4, S-2)

3.1   Janis, Maija, and their mother were eating a cake. Janis ate 1/2 of the cake. Maija ate 1/4 of the cake. Their mother ate 1/4 of the cake. How much of the cake is left? (Adapted from TIMSS gr. 8, P-14)

3.1   Sam, who is 6 years old, likes vanilla ice cream with his apple pie. Sam said that 1/3 of an apple pie is less than 1/4 of the same pie. Is Sam correct in his estimate? (Adapted from TIMSS gr. 4, V-1)

Measurement and Geometry

1.2   Make an outline of your hand with your fingers together on a piece of grid paper. Assuming that each grid is 1 cm^2, what is roughly the area of your hand?

Grade Four

Algebra and Functions

1.4 Maria and her sister, Louisa, leave home at the same time and ride their bicycles to school 9 kilometers away. Maria rides at a rate of 3 kilometers in 10 minutes. How long will it take her to get to school? Louisa rides at a rate of 1 kilometer in 3 minutes. How long will it take her to get to school? Who arrives first? (Adapted from TIMSS gr. 4, U-3)

1.5 My plane was supposed to leave San Francisco at 8:42 a.m. and arrive in Los Angeles at 9:55 a.m. But it started 11 minutes late, and to make up for lost time, the pilot increased the speed and shortened the flight time to 58 minutes. What time did I arrive in Los Angeles? (This problem also applies to Mathematical Reasoning Standard 1.1.)

2.1 What is the remainder when 1,200,354,003 is divided by 5?

Measurement and Geometry

1.2, 1.4 Given 12 square tiles, all the same size, describe all the rectangles you can that use all the tiles. Find the perimeter of each rectangle.

3.7 Assume that the sum of the length of any two sides of a triangle is greater than the length of the third side. If the lengths of the sides of a triangle are required to be whole numbers, how many such triangles are there with a perimeter of 14? List all of them.

Statistics, Data Analysis, and Probability

1.1 If six people enter a room and shake hands with each other once, how many handshakes occur?

Grade Five

Number Sense

1.2 Change to decimals: $\frac{17}{1,000}$, $\frac{3}{20}$, 6%, 35½

1.2 Change to fractions: 0.03, 1.111, 8%, 21

1.2 Change to percents: 0.07, 0.165, $\frac{17}{20}$, $\frac{1}{8}$

1.2 6 is what % of 25?

1.2 What is 15% of 44?

1.2 30 is 20% of what?

1.2 Betty paid $23.60 for an item that was reduced by 20%.
1. What was the original price?
2. If the original price was reduced by 25%, what is the sale price?

1.4 Write as a product of primes using exponents (use factor trees or equivalents):

18, 48, 100

1.3, 1.4 What is the largest square of a whole number that divides 48? What is the largest cube of a whole number that divides 48?

1.5 Arrange in order from smallest to largest:

¾, 25%, 0.3, 2½, 0.295

1.5 Do the following problems mentally.

$$\frac{(9185}{(2117} \times \frac{12)}{13)} + \frac{9185}{2117} = ?$$

$$\frac{(9185}{(13} \times \frac{12\ )}{2117)} + \frac{9185}{2117} = ?$$

2.1 Find the average of 6.81, 7, 5.2 and round the answer to the nearest hundredth.

2.1 Evaluate 0.25 (3 − 0.75).

Measurement and Geometry

1.0 Find the areas (dimensions are in cm):

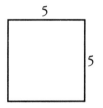 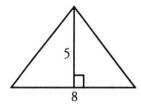 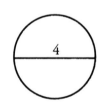

1.1 How many segments x will fit on the perimeter of the square?

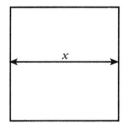

1.3, 1.4 Determine the volume of a rectangular solid with base 65 cm, height 70 cm, and width 50 cm. For the same rectangular solid, determine its surface area. (Make sure that your answer is expressed in the correct units.)

1.4 Identify the relevant dimension as length, area, or volume:

1. The perimeter of a triangle
2. The capacity of a barrel
3. The capacity of a box
4. The amount of sod needed to cover a football field
5. The number of bricks needed to pave a path
6. The height of a tree

2.1 Explain how to construct various complex geometric figures by using a ruler and compass; e.g., an equilateral triangle, a regular hexagon, a line passing through a given point and perpendicular to a given line.

2.2 Find the third angle of a triangle if you know that one angle is 60° and the second angle is 20°.

Statistics, Data Analysis, and Probability

1.2. Draw a circle graph to display the following data: A certain municipal district spends 6 million dollars per year—$2,507,000 on education, $1,493,000 for public safety, $471,000 for libraries, $536,000 for road maintenance, and $993,000 for miscellaneous expenses. (This problem also applies to Number Sense Standards 1.1 and 1.2.)

Grade Six

Number Sense

1.2, 1.3 Complete the following statements:

1. If 3 ft. = 1 yd., then 7 ft. = ? yd.
2. If 32 oz. = 1 qt., then 6.7 qt. = ? oz.

1.2, 1.3 In a lemonade punch, the ratio of lemonade to soda pop is 2:3. If there are 24 gallons of punch, how much lemonade is needed?

2.1 Find the sum $\frac{5}{6} + \frac{3}{10}$.

2.3, 2.4 Write the following as an integer over a whole number:

$8, -6, 4\frac{1}{2}, -1\frac{1}{5}, 0, 0.013, -1.5$

2.4
1. Find the least common multiple of 6 and 10 (count by sixes until you come to a multiple of 10).
2. List the first 20 multiples of 6.
3. List the first 20 multiples of 10.
4. List the multiples that 6 and 10 have in common above 120.

2.4
1. Make a sieve of Eratosthenes up to 100.
2. Find the greatest common factor of 18 and 30 (list all factors of 18 until you come to a factor of 30).
3. Reduce $\dfrac{18}{30}$.

Algebra and Functions

1.2 Moe was paid $7 per hour and earned $80.50. How many hours did Moe work?

1.2 Write the following in symbolic notation using n to represent the number:
1. A number increased by 33
2. The product of a number and (−7)
3. $8\frac{1}{2}$ decreased by some number
4. The square of some number which is then divided by 7
5. The sum of some number and $\frac{1}{3}$ which is then increased by the third power of the same number

1.2, 3.1 A rectangle is constructed with 8 feet of string. Suppose that one side is $1\frac{4}{14}$ feet long. What is the length of the other side?

1.3 True or false?
$(25 + 16) \times 6 = 25 + 16 + 6$.

Measurement and Geometry

1.2 How many segments x will fit on the circumference of the circle?

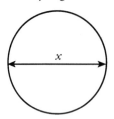

1.1, 1.3 Use the formula $\pi r^2 h$ for the volume of a right circular cylinder. What is the ratio of the volume of such a cylinder to the volume of one having half the height but the same radius? What is the ratio of the volume of such a cylinder to the volume of one having the same height but half the radius? (This problem also applies to Number Sense Standard 1.2.)

2.2   Line *L* is parallel to line *M*. Find the missing angles.

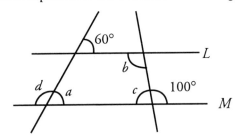

2.1   Line *L* is parallel to line *M*. Line *P* is perpendicular to *L* and *M*. Name the following. If none can be named, leave the space blank.

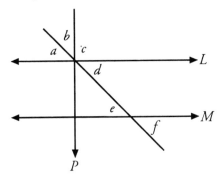

1. Complementary _____
2. Supplementary _____
3. Vertical _____
4. Alternate interior _____
5. Corresponding _____
6. Acute _____
7. Right _____
8. Obtuse _____

Statistics, Data Analysis, and Probability

2.2, 2.5   50 red marbles are placed in a box containing an unknown number of green marbles. The box is thoroughly mixed, and 50 marbles are taken out. Ten of those marbles are red. Does this imply that the number of green marbles was 200?

3.1, 3.4   Make a tree diagram of all the possible outcomes of four successive coin tosses. How many paths in the tree represent two heads and two tails? Suppose the coin is weighted so that there is a 60% probability of heads with each coin toss. What is the probability of two heads and two tails?

Grade Seven

Number Sense

1.1 Convert to scientific notation; compute and express your answer in scientific notation and in decimal notation.

1. $\dfrac{(350,000)(0.0049)}{.25}$

2. $\dfrac{(0.000042)(0.0063)}{((140,000)(70,000)(0.18))}$

1.6 Peter was interested in buying a basketball. By the time he saved enough money, everything in the sporting goods store had been marked up by 15%. Two weeks later, however, the same store had a sale, and everything was sold at a 15% discount. Peter immediately bought the ball, figuring that he was paying even less than before the prices were raised. Was he mistaken?

1.7 What will be the monthly payments on a loan of $50,000 at 12% annual interest so that it will be paid off at the end of 10 years? How much total interest will have been paid? Do the same problem with 8% annual interest over 10 years. Do the same problem with 10% annual interest over 15 years. (Use calculators.)

2.2 Reduce $\dfrac{910}{1,859}$.

2.2 Subtract and reduce to lowest terms:

$$\left(\dfrac{81}{143}\right) - \left(\dfrac{7}{208}\right)$$

(For clarification see the discussion in Appendix A on the addition and subtraction of fractions.)

2.4 Determine without a calculator which is bigger: $\sqrt{291}$ or 17?

2.5 Consider two numbers A and B on the number line. Determine which is larger: the distance between A and B or the distance between $|A|$ and $|B|$? Always? Sometimes? Never?

Algebra and Functions

1.1 Gabriel bought a CD player, listed at a, at a 20% discount; he also had to pay an 8% sales tax. After three months he decided that its sound quality was not good enough for his taste, and he sold it in the secondhand market for b, which was 65% of what he paid originally. Express b as a function of a.

1.1, 4.2 A car goes 45 mph and travels 200 miles. How many hours will it take for the car to reach its destination?

1.1, 4.2 A plane flying at 450 mph leaves San Francisco. One-half hour later a second plane flying at 600 mph leaves, flying in the same direction. How long will it take the second plane to catch the first? How far from San Francisco will this event happen?

1.5 Water is poured at a constant rate into a flask shaped like the one in the illustration that follows. Draw a graph of the water level in the flask as a function of time.

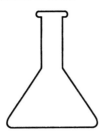

2.1, 2.2 Simplify as a nonfraction:

1. $\dfrac{x^5}{x^3}$

2. $\dfrac{x^3}{x^5}$

3. $\dfrac{x^5}{x^5}$

4. $\dfrac{\left(42a^5 b^3\right)}{\left(14a^2 b^9\right)}$

5. $\dfrac{x^{-8}}{x^{-7}}$

6. $\dfrac{\left(a^7 b^{-3} c^9\right)}{\left(a^4 b^{-3} c^{10}\right)}$

3.1, 3.2 Write the equation of the surface area of a cube of side length x. Graph the surface area as a function of x.

3.1 The amount of paint needed to paint over a surface is directly proportional to the area of the surface. If 2 quarts of paint are needed to paint a square with a side of 3 ft., how many quarts must be purchased to paint a square whose side is 4 ft. 6 in. long?

3.4 What is the slope of the straight line which is the graph of the function expressing the length of a semicircle as a function of the radius?

4.1 Becky and her sister have some money. The ratio of their money is 3:1. When Becky gives $5 to her sister, their ratio will be 2:1. How much money does Becky have? (World Math Challenge 1995)

4.2 Three people set out on a car race to see who would be the first to get to town T and back. Anne maintained a steady speed of 80 mph throughout the race. Lee averaged 90 mph on the way out, but he could manage only an average of 70 mph on the way back. Javier started slowly and averaged 65 mph during the first third of the race, but he increased his speed to 85 mph in the second third and finished with a blazing 100 mph in the last third. Who won? (*Note:* This is a difficult problem that would be particularly good for advanced students.)

Measurement and Geometry

1.1 Know the following approximations:

1. 1 meter ≈ 1 yard (baseball bat)
2. 1 cm ≈ ½ inch (width of a fingernail)
3. 1 km ≈ .6 miles
4. 1 kg ≈ 2.2 lbs. (a textbook)
5. 1 liter ≈ 1 quart
6. 1 gram ≈ (1 paper clip)
7. 1 mm ≈ (thickness of a dime)

1.3 A bucket is put under two faucets. If one faucet is turned on alone, the bucket will be filled in 3 minutes; if the other is turned on, the bucket will be filled in 2 minutes. If both are turned on, how many seconds will it take to fill the bucket?

2.1 Compute the area and perimeter of a regular hexagon inscribed in a circle of radius 2.

2.1 Compute the volume and surface area of a square-based pyramid whose lateral faces are equilateral triangles with each side equal to 4.

3.2 Determine the vertices of a triangle, whose vertices were originally at (1, 2), (−3, 0), and (−1, 5), after it is translated 2 units to the right and 1 unit down and then reflected across the graph of $y = (1/2)x - 3$.

3.3 What is the distance from the center of a circle of radius 3 to a chord of length 5 cm?

3.3   Find the missing angles and arcs. (∠B is 10°.)

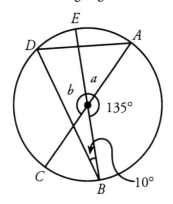

1. Minor $\overarc{AB}$
2. Angle a
3. Major $\overarc{BA}$
4. Angle b
5. $\overarc{EC}$
6. $\overarc{CB}$
7. $\overarc{ED}$
8. $\overarc{DC}$

Grades Eight Through Twelve

Algebra I

1.0, 24.0   Prove or give a counterexample: The average of two rational numbers is a rational number.

2.0   I start with a number and apply a four-step process: I (1) add 13; (2) multiply by 2; (3) take the square root; and (4) take the reciprocal. The result is 1/4. What number did I start with? If I start with the number x, write a formula that gives the result of the four-step process.

2.0   What must be true about a real number x if $x = \sqrt{x^2}$?

2.0   Write as a power of x: $\dfrac{\sqrt{x}}{x \cdot \sqrt[3]{x}}$.

2.0   Solve for x: $\left|x^3\right| = \dfrac{1}{2\sqrt{2}}$.

3.0   Solve for x: $3|x| + 2 = 14$.

3.0   Express the solution using interval notation: $|x+1| \geq 2$.

Sketch the interval in the real number line that is the solution for $|x-5| < 2$.

4.0   Expand out and simplify $2(3x+1) - 8x$.

4.0   Solve for x in each case:

$$5x - 2 \leq -3(x+1) + 2$$
$$2 - (2 - (3x+1) + 1) = 3(x-2) + x$$
$$\frac{3}{x-2} = \frac{4}{x+5}$$

5.0   To compute the deduction that you can take on your federal tax return for medical expenses, you must deduct 7.5% of your adjusted gross income from your actual medical expenses. If your actual medical expenses are $1,600 and your deduction is less than $100, what can you conclude about your adjusted gross income? (CERT 1997)

5.0   Joe is asked to pick a number less than 100, and Moe is asked to guess it. Joe picks 63. Write an inequality that says that Moe's guess is within 15, inclusive, of the number Joe has in mind. Solve this inequality to find the range of possibilities for Moe's guess.

5.0   Four more than three-fifths of a number is 24. Find the number.

5.0   Luis was thinking of a number. If he multiplied the number by 7, subtracted 11, added 5 times the original number, added −3, and then subtracted twice the original number, the result was 36. Use this information to write an equation that the number satisfies and then solve the equation.

6.0   The cost of a party at a local club is $875 for 20 people and $1,100 for 30 people. Assume that the cost is a linear function of the number of people. Write an equation for this function. Sketch its graph. How much would a party for 26 people cost? Explain and interpret the slope term in your equation. (CERT forthcoming)

6.0   Graph $2x - 3y = 4$. Where does the line intersect the x-axis? Where does the line intersect the y-axis? What is the slope?

6.0   Sketch the region in the x-y plane that satisfies both of the following inequalities: $y < 3x + 1$, $2x + 3y + 8 > 0$

6.0, 7.0   Find an equation for the line that passes through (2, 5) and (−3, 1). Where does the line intersect the x-axis? Where does the line intersect the y-axis? What is the slope?

6.0, 7.0   Find an equation for the line that passes through (5, 3) and (5, −2). Where does the line intersect the x-axis? Where does the line intersect the y-axis? What is the slope?

7.0	The weight of a pitcher of water is a linear function of the depth of the water in the pitcher. When there are 2 inches of water in the pitcher, it weighs 2 lbs.; and when there are 8 inches of water in the pitcher, it weighs 5 lbs. Find a formula for the weight of the pitcher as a function of the depth of the water.
7.0	Find an equation for the line that passes through (−2, 5) and has slope −2/3.
8.0	Find the equation of the line that is perpendicular to the line through (2, 7) and (−1, 3) and passes through the x-intercept of that line.
8.0	Are the following two lines perpendicular, parallel, or neither? $$2x + 3y = 5$$ $$3x + 2y - 1 = 0$$
8.0	If the line through (1, 3) and (a, 9) is parallel to $3x - 5y = 2$, what is a?
9.0	Line 1 has equation $3x + 2y = 3$, and line 2 has equation $-2x + y = 5$. Find the point of intersection of the two lines.
9.0	Sketch a graph of the values of x and y that satisfy both of the inequalities: $$3x + 2y \geq 3$$ $$-2x + y \leq 5$$
10.0	The volume of a rectangular prism with a triangular base is $36m^3 - 72m^2 + 29m - 3$. Assume that the height of the prism is $3m - 1$ and the height of the triangle is $6m - 1$. What is the base of the triangle?
10.0	Simplify $[(3b^2 - 2b + 4) - (b^2 + 5b - 2)](b + 2)$.
11.0	Solve for x: $\dfrac{x^2 - 4}{x - 2} + x^2 - 4 = 0$.
12.0	Reduce to lowest terms: $\dfrac{x^3 + x^2 - 6x}{x^2 + 13x + 30}$.
12.0	Solve for x: $\dfrac{3}{x - 1} + \dfrac{10}{x^2 - 2x + 1} = 4$.
13.0	Solve for x: $\dfrac{x + 2}{x - 3} \cdot \dfrac{x^2 + 5x - 24}{x - 6} + 3 = 0$.
14.0	Where does the graph of $f(x) = \dfrac{x^3 + 2x^2 - 15x}{x + 1}$ intersect the x-axis?
15.0	Mary drove to work on Thursday at 40 mph and arrived one minute late. She left at the same time on Friday, drove at 45 mph, and arrived one minute early. How far does Mary drive to work? (CERT forthcoming)

15.0 Suppose that peanuts cost $.40/lb. and cashews cost $.72/lb. How many pounds of each should be used to make an 80 lb. mixture that costs $.48/lb.?

16.0 The following points lie on the graph of a relation between x and y:

(0, 0), (−2, 3), (3, −2), (2, 3), (−3, −3), (2, −2)

Can y be a function of x? Explain. Can x be a function of y? Explain.

17.0 Determine the domain of the function $f(x) = \sqrt{|x| - 6}$.

17.0 Determine the range of the function g whose graph is shown below.

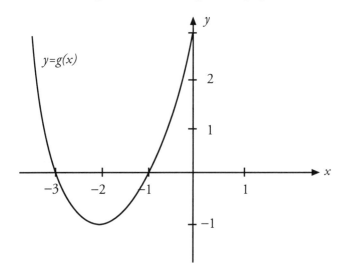

18.0 Does the equation $x^2 + y^2 = 1$ determine y as a function of x? Explain.

20.0 Solve for x: $2x^2 - 3x - 5 = 0$.

20.0 Let $f(x) = ax^2 + bx + c$. Suppose that $b^2 - 4ac > 0$. Use the quadratic formula to show that f has two roots.

22.0 At how many points does the graph of $g(x) = 2x^2 - x + 1$ intersect the x-axis?

23.0 A ball is launched straight up into the air from the ground at a rate of 64 feet per second. Its height h above the ground (in feet) after t seconds is $h = 64t - 16t^2$.

How high is the ball after 1 second? When is the ball 64 feet high? For what values of t is $h = 0$? What events do these represent in the flight of the ball? (CERT 1997)

23.0 The braking distance of a car (how far it travels after the brakes are applied until it comes to a stop) is proportional to the square of its speed. Write a formula expressing this relationship and explain the meaning of each term in the formula. If a car traveling at 50 mph has a breaking distance of 105 feet, then what would its braking distance be if it were traveling 60 mph? (ICAS 1997)

24.0   Provide numbers to show how the following can be false and if possible describe when it is true:

$\sqrt{a^2 + b^2} < a + b$ whenever $a \geq 0$ and $b \geq 0$

(CERT 1997)

25.0   Suppose that 9 is a factor of xy, where x and y are counting numbers. At least one of the following is true. Which of the following statements are necessarily true? Explain why.

  a. 9 must be a factor of x or of y.

  b. 3 must be a factor of x or of y.

  c. 3 must be a factor of x and of y.

(CERT forthcoming)

25.0   A problem is given, to find all solutions to the equation $(2x + 4)^2 = (x + 1)^2$. Comment on any errors in the following proposed solutions:

$$(2x + 4)^2 = (x + 1)^2$$

Take the square root of both sides to find $2x + 4 = x + 1$

Subtract x and 4 from both sides to obtain $2x + 4 - x - 4 = x + 1 - x - 4$

Simplify to conclude $x = -3$

Geometry

3.0   Prove or disprove: Any two right triangles with the same hypotenuse have the same area.

3.0   True or false? A quadrilateral is a rectangle only if it is a square.

3.0   Suppose that all triangles that satisfy property A are right triangles. Is the following statement true or false? A triangle that does not satisfy the Pythagorean theorem does not satisfy property A.

4.0   Suppose that triangle PRS is isosceles, with $\overline{RP} = \overline{PS}$. Show that if the segment PQ bisects the $\angle RPS$, then $\overline{RQ} = \overline{QS}$.

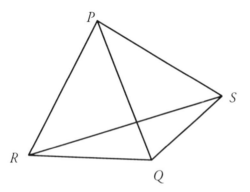

4.0   Suppose that R and S are points on a circle. Prove that the perpendicular bisector of the line segment RS passes through the center of the circle.

5.0   In the figure shown below, the area of the shaded right triangle is 6. Find the distance between the parallel lines, L_1 and L_2. Explain your reasoning. (CERT forthcoming)

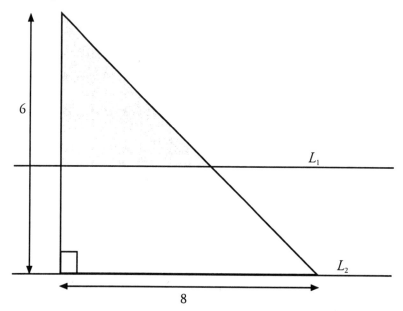

6.0   Using a geometric diagram, show that for any positive numbers a and b, $\sqrt{a^2 + b^2} < a + b$.

7.0, 4.0   On the following diagram, with distances as shown, prove that if $x = y$, then the lines L and M are parallel:

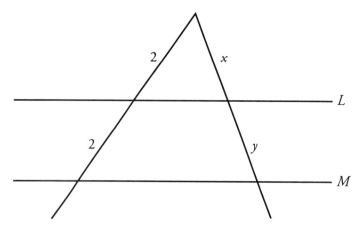

7.0   Prove that if a diagonal of a parallelogram bisects an angle of a parallelogram, then the parallelogram is a rhombus.

7.0   Prove that if the base angles of a trapezoid are congruent, then the trapezoid is isosceles.

8.0   A string is wound symmetrically around a circular rod. The string goes exactly one time around the rod. The circumference of the rod is 4 cm, and its length is 12 cm. Find the length of the string. What is the length of the string if it goes exactly four times around the rod? (Adapted from TIMSS)

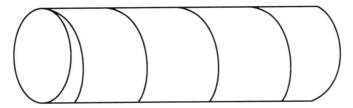

9.0   A sphere of radius 1 is inscribed in a cylinder. Find the volume of the cylinder.

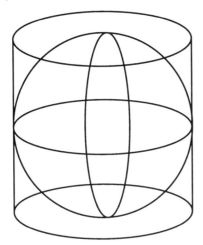

9.0   A right prism with a 4-inch height has a regular hexagonal base. The prism has a volume of 144^3 inches. Find the surface area of the prism.

10.0 A trapezoid with bases of length 12 and 16 is inscribed in a circle of radius 10. The center of the circle lies inside the trapezoid. Find the area of the trapezoid.

11.0 Brighto soap powder is packed in cube-shaped cartons. A carton measures 10 cm on each side. The company decides to increase the length of each edge of the carton by 10 percent. How much does the volume increase? (TIMSS)

12.0 A regular polygon has exterior angles, each measuring 10 degrees. How many sides does the polygon have?

13.0 Prove that if the diagonals of a quadrilateral bisect each other, then the quadrilateral is a parallelogram.

15.0 Find the length of the side labeled C in the figure shown below:

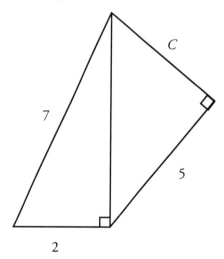

15.0 The bottom of a rectangular box is a rectangle with a diagonal whose length is $4\sqrt{3}$ inches. The height of the box is 4 inches. Find the length of a diagonal of the box.

16.0 Given a circle, use an unmarked straightedge and a compass to find the center of the circle.

17.0 The vertices of a triangle PQR are the points P(1, 2), Q(4, 6), and R(−4, 12). Which one of the following statements about triangle PQR is true?

1. PQR is a right triangle with right $\angle P$.
2. PQR is a right triangle with right $\angle Q$.
3. PQR is a right triangle with right $\angle R$.
4. PQR is not a right triangle. (TIMSS)

18.0 Shown below is a semicircle of radius 1 and center C. Express the unknown length D in terms of the angle a:

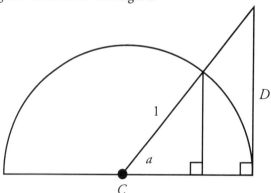

18.0 If α is an acute angle and $\cos \alpha = \frac{1}{3}$, find $\tan \alpha$.

19.0 Find the length of side *C* below, if ∠*a* measures 70 degrees:

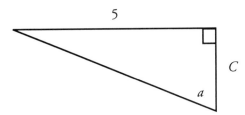

20.0 Each side of the regular hexagon *ABCDEF* is 10 cm long. What is the length of the diagonal *AC*? (TIMSS)

20.0 Express the perimeter of the trapezoid *ABCD* in the simplest exact form. Angle *DAB* measures 30 degrees, and angle *ABC* measures 60 degrees.

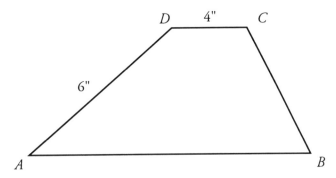

21.0 Two circles with centers *A* and *B*, as shown below, have radii of 7 cm and 10 cm, respectively. If the length of the common chord *PQ* is 8 cm, what is the length of *AB*? Show all your work. (TIMSS)

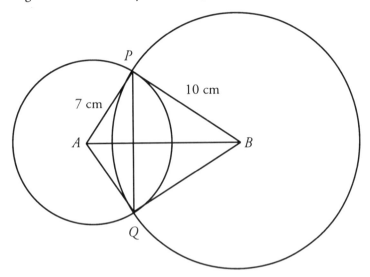

22.0 A translation maps *A* (2, −3) onto *A'* (−3, −5). Under the same translation, find the coordinates of *B'*, the image of *B* (1, 4). (TIMSS)

22.0 Which response listed below applies to the statement that follows? The rectangle labeled Q cannot be obtained from the rectangle P by means of:

1. Reflection (about an axis in the plane of the page)
2. Rotation (in the plane of the page)
3. Translation
4. Translation followed by a reflection (TIMSS)

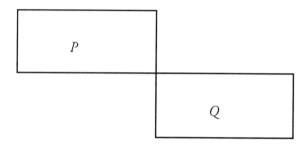

Algebra II

1.0 Express the solution using interval notation:
$$|2x - 3| > 4$$

1.0 Sketch the interval in the real number line that is the solution for:
$$\frac{|x - 3|}{2} > 5$$

2.0 Solve the system of linear equations:
$$x + 2y = 0$$
$$x + z = -1$$
$$y - z = 2$$

4.0 Simplify $\dfrac{x^3 - y^3}{x^2 - y^2}$.

4.0 Simplify $\dfrac{\sqrt{x} + y}{x - y^2}$.

5.0 Locate all complex solutions to $z^2 + 4$ in the complex plane.

6.0 Write in the form $a + bi$, where i is a square root of -1:
$$\frac{(3 - 2i)^2}{2 + i}$$

8.0 Find all solutions to the equation $x^2 + 5x + 8 = 0$.

9.0 The function $f(x) = (x - b)^2 + c$ is graphed below. Use this information to identify the constants b and c.

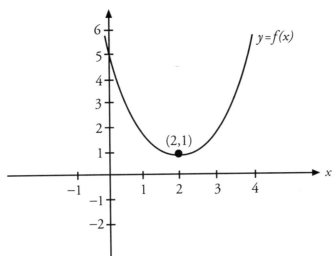

10.0 Graph the function $f(x) = 2(x + 3)^2 - 4$ and determine the minimum value for the function.

10.0 Find the vertex for the graph of $f(x) = 3x^2 - 12x + 4$.

11.0 Solve for x in each of the following and explain each step:

$$\log_3(x+1) - \log_3 x = 1$$
$$\log_{\sqrt{b}} 7 = \log_b x$$

12.0 Scientists have observed that living matter contains, in addition to carbon, C12, a fixed percentage of a radioactive isotope of carbon, C14. When the living material dies, the amount of C12 present remains constant, but the amount of C14 decreases exponentially with a half life of 5,550 years. In 1965 the charcoal from cooking pits found at a site in Newfoundland used by Vikings was analyzed, and the percentage of C14 remaining was found to be 88.6. What was the approximate date of this Viking settlement? (ICAS 1997)

13.0 Simplify to find exact numerical values for:

$$\log_{\sqrt{b}}(b^2)$$
$$b^{3\log_b 2 - \log_b 5}$$

14.0 Write as a single logarithm $\dfrac{\log_3 7}{\log_3 5}$.

15.0 Is the following true for all real numbers x, for some real numbers x, or for no real numbers x?

$$\dfrac{\sqrt{(1-x^2)^2}}{1-x} = 1 + x$$

16.0 If $xy = 1$ and x is greater than 0, which of the following statements is true?

1. When x is greater than 1, y is negative.
2. When x is greater than 1, y is greater than 1.
3. When x is less than 1, y is less than 1.
4. As x increases, y increases.
5. As x increases, y decreases. (TIMSS)

17.0 Write in standard form the conic section whose equation is given by $4x^2 - 8x - y^2 + 4y = 4$ to determine whether it is a parabola, a hyperbola, or an ellipse.

18.0 An examination consists of 13 questions. A student must answer only one of the first two questions and only nine of the remaining ones. How many choices of questions does the student have? (Adapted from TIMSS)

19.0 A lottery will be held to determine which three members of a club will attend the state convention. This club has 12 members, 5 of whom are women. What is the probability that none of the representatives of the club will be women?

20.0 Determine the middle term in the binomial expansion of $\left(x - \dfrac{2}{x}\right)^{10}$. (ICAS 1997)

21.0 Use mathematical induction to show that
$$1 + 2 + 3 + 4 + \cdots + n = \dfrac{n(n+1)}{2}$$

22.0 Find the sum of the following infinite series:
$$\dfrac{3}{5} + \dfrac{9}{25} + \dfrac{27}{125} + \dfrac{81}{625} + \cdots$$

24.0 Sketch a graph of a function g that satisfies the following conditions: g does not have an inverse function, $g(x) < x$ for all x, and $g(2) > 0$.

Trigonometry

1.0 Express in degrees:

$\pi/5$ radians

1/8 revolution

1.0   Find the indicated angle B in radians, if C is the center of the circle:

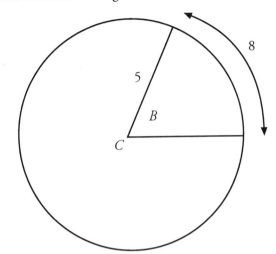

2.0   Graph the functions $f(x) = \sin x$ and $g(x) = \cos x$, where x is measured in radians, for x between 0 and 2π. Identify the points of intersection of the two graphs.

3.0   Prove that $\sec^2 x + \csc^2 x = \sec^2 x \cdot \csc^2 x$.

5.0   Use the definition of $f(x) = \tan(x)$ to determine the domain of f.

6.0   Identify all vertical asymptotes to the graph of $g(x) = \sec x$.

7.0   A line with positive slope makes an angle of 1 radian with the positive x-axis at the point (3, 0). Find an exact equation for this line.

8.0   If $\tan(x) = \tan(\pi/5)$ and $3\pi < x < 4\pi$, find x.

8.0   Graph $f(x) = \sin x$ and the principal value of $g(x) = \sin^{-1} x$ on the same axes. Write a description of the relationship between the two graphs.

9.0   Find an angle α between 0 and $-\pi$ for which $\cos(\alpha) = -1/2$.

11.0   Solve for θ, where $0 < \theta < 2\pi$: $(\cos \theta)(\sin 2\theta) - 2\sin \theta + 2 = 0$.

12.0   Find the measure of the angle a in the triangle below:

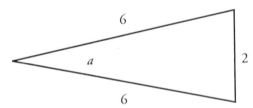

13.0 Solve for the distance c on the triangle shown below, if the angle A is 30°:

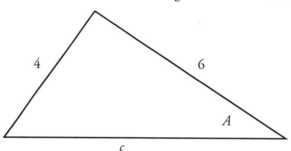

14.0 Find the area of the triangle shown below if the angle B measures 20 degrees:

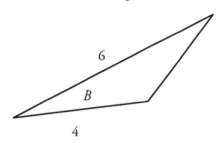

15.0 Find all representations in polar coordinates of the point whose rectangular coordinates are $(2\sqrt{3}, -2)$.

17.0 Represent $i + 1$ in polar form. Use this to compute $(i + 1)^2$.

18.0 Find all square roots of i.

19.0 A person holds one end of a rope that runs through a pulley and has a weight attached to the other end. The section of rope between the person and the pulley is 20 feet long; the section of rope between the pulley and the weight is 10 feet long. The rope bends through an angle of 35 degrees in the pulley. How far is the person from the weight?

19.0 How long does it take for a minute hand on a clock to pass through 1.5 radians?

Mathematical Analysis

1.0 Find any points of intersection (first in polar coordinates and then in rectangular coordinates) of the graphs of $r = 1 + \sin\theta$ and the circle of radius 3/2 centered about the origin. Verify your solutions by graphing the curves. Find any points of intersection (first in polar coordinates and then in rectangular coordinates) of the graphs of $r = 1 + \sin\theta$ and the line with slope 1 that passes through the origin. Verify your solutions by graphing the curves. (ICAS 1997)

2.0 Compute $\left(\dfrac{1}{2} - \dfrac{\sqrt{3}}{2}i\right)^{11}$.

5.0   Consider the locus of points in the plane whose distance to (0, 1) is twice its distance from (0, −2). Identify this conic section and find its equation in standard form.

6.0   Sketch the graph of $f(x) = \dfrac{x}{x^2 - 4}$, showing all asymptotes.

7.0   Sketch a graph of the curve determined by the equations:

$$x = \cos(t^2) + 1$$
$$y = \sin(t^2) \quad \text{for } 0 < t < 5$$

and find another set of parametric equations that describe the same curve.

Probability and Statistics

1.0   A warning system installation consists of two independent alarms having probabilities of 0.95 and 0.90, respectively, of operating in an emergency. Find the probability that at least one alarm operates in an emergency. (Adapted from TIMSS)

1.0   Arlene and her friend want to buy tickets to an upcoming concert, but tickets are difficult to obtain. Each ticket outlet will have its own lottery so that everyone who is in line at a particular outlet to buy tickets when they go on sale has an equal chance of purchasing them. Arlene goes to a ticket outlet where she estimates that her chance of being able to buy tickets is 1/2. Her friend goes to another outlet, where Arlene thinks that her chance of being able to buy tickets is 1/3.

1. What is the probability that both Arlene and her friend are able to buy tickets?
2. What is the probability that neither Arlene nor her friend is able to buy tickets?
3. What is the probability that at least one of the two friends is able to buy tickets? (CERT 1997)

3.0   A random variable X has the following distribution:

x	−1	0	2	3	4
$P(X=x)$	.1	.3	.2	.1	.3

Find: $P(X > 1)$ $P(X^2 < 2)$

3.0   A fund-raising group sells 1,000 raffle tickets at $5 each. The first prize is a $1,800 computer. The second prize is a $500 camera, and the third prize is $300 in cash. What is the expected value of a raffle ticket? (ICAS 1997)

3.0   Carla has made an investment of $100. She understands that there is a 50% chance that after a year, her investment will have grown to exactly $150. There is a 20% chance that she will double her money in that year, but there is also a 30% chance that she will lose the entire investment. What is the expected value of her investment after a year? (CERT 1997)

4.0   You are playing a game in which the probability that you will win is 1/3, and the probability that you will lose or play to a tie is 2/3. If you play this game 8 times, what is the probability that you will win exactly 3 times?

5.0   Suppose that X is a normally distributed random variable with mean μ. Find $P(X < \mu)$.

Advanced Placement Probability and Statistics

1.0   I roll two standard fair dice and look at the numbers showing on the top sides of the two dice. Let A be the event that the sum of the two numbers showing is greater than 5. Let B be the event that neither die is showing a 1 or a 6. Are events A and B independent?

5.0   Suppose that X is a discrete random variable and that X has the following distribution:

x	−1	0	2	5
$P(X=x)$	1/4	1/8	1/2	1/8

Compute the mean for X.

6.0   Suppose that X is a discrete random variable and that X has the following distribution:

x	−1	0	2	5
$P(X=x)$	1/4	1/8	1/2	1/8

Compute the variance for X.

9.0   Suppose that a new medical treatment is reported to be successful for 80% of patients. What is the probability that in a sample of 100 patients, 75 or more will find the treatment successful?

12.0 A teacher claims that quiz scores for students are indicative of their test scores. You sample 6 students from this teacher's class and find the following quiz and test scores:

Quiz scores	7	2	9	6	9	5
Test scores	85	60	80	70	85	80

Draw a scatter plot for these data, with the quiz scores on the horizontal axis and test scores on the vertical axis. Find the line that best fits these data by using least squares and graph the line along with your scatter plot.

13.0   In the preceding example concerning quiz scores and test scores, by using the graph alone, what can you say about the correlation and coefficient? Suppose that 4 more data points are collected and that the best fit line remains approximately the same for the combined data, but that the correlation coefficient now is closer to 1 than it was for just the 6 data points. What can you say about the placement of the 4 additional data points?

16.0   Suppose that it is known that the average lifetime of a particular brand of light bulb is 1,000 hours, with a standard deviation of 90 hours. You sampled 20 of these bulbs and computed that their lifetimes averaged 900 hours, with a sample standard deviation of 120 hours. If you sample another 20 bulbs and combine your data, what is most likely to occur to the average lifetimes for the 40 bulbs and to the sample standard deviation for the 40 bulbs?

17.0   Suppose that the number of cars passing a certain bridge on a freeway during one-minute intervals is normally distributed. Suppose that 61 one-minute observations are randomly made. The average number of cars passing the bridge in a minute, over the 61 observations, is 31. The sample variance is 25. Find a 95% confidence interval for the average number of cars passing the bridge per minute. For a margin of error of 1, with 95% confidence, how large a sample size would be needed?

18.0   In the situation described previously, if you were testing the hypothesis that the average number of cars per minute traveling over the bridge is more than 30, what P-value would you attach to the data that were collected?

Calculus

1.0   Without using a graphing calculator, evaluate $\lim_{x \to \infty} \arctan x$. Explain what this limit should mean about the graph of the arctangent function and then verify this limit on a graphing calculator.

1.0   Using the graph of f shown below, estimate:

$$\lim_{x \to 0} f(x) \qquad \lim_{x \to 1^-} f(x) \qquad \lim_{x \to 1^+} f(x) \qquad \lim_{x \to 2^-} f(x)$$

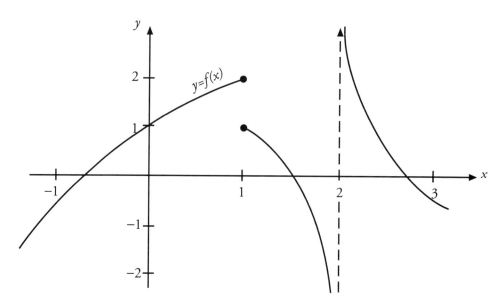

1.0   Using the formal definition of limit, show that $\lim_{x \to 2}(3x+1) = 7$.

1.0   Using the formal definition of limit, show that $\lim_{x \to 0} \dfrac{x}{|x|}$ does not exist.

2.0   Using the formal definitions of continuity and limit, show that $f(x) = .5x + 4$ is continuous.

3.0   Use the Intermediate Value theorem to assert that the equation $4^x = x + 5$ has a solution.

3.0   Give an example that demonstrates that the conclusions of the Intermediate Value theorem need not hold for a function that is not continuous.

3.0   Must $f(x) = \dfrac{|x|}{x+3}$ have a maximum and a minimum value on the interval $[-1, 3]$? Explain.

4.0   Using the definition of derivative, find the derivative of $f(x) = \sqrt{x+1}$.

4.0   Differentiate:

$$f(x) = |\sin x|$$

$$g(x) = \dfrac{1 + \ln x}{e^x}$$

$$h(x) = \arctan(x)$$

4.0 Using the graph of f shown below, estimate:

$f'(1)$ $f'(0)$ $\lim_{x \to \infty} f'(x)$

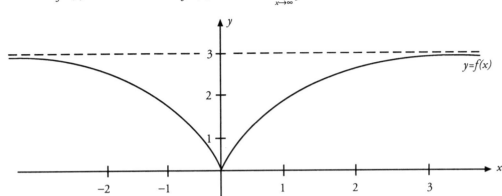

4.0 Find the value of $\lim_{h \to 0} \dfrac{\sqrt{2+h} - \sqrt{2}}{h}$. (Adapted from TIMSS)

4.0 A brush fire spreads so that after t hours, $80t - 20t^2$ acres are burning. What is the rate of growth of the acreage that is burning after 90 minutes?

4.0 According to Newton's law of gravitation, a particle of mass m attracts a particle of mass M with a force whose magnitude is $F = \dfrac{GmM}{r^2}$, where G is the gravitational constant and r is the distance between the two particles. For particles that are in motion, find the rate of change of F with respect to r.

5.0 Differentiate:

$$g(x) = \ln(x + e^{\cos(x)})$$

$$k(x) = e^{\sqrt{\ln(5-x)}}$$

5.0 Use the facts that $f(x) = \log_2 x$ and $g(x) = 2^x$ are inverse functions and that $g'(x) = 2^x \ln 2$ to find the derivative of $f(x)$.

6.0 A wheel of radius 1 rolls on a straight line (the x-axis) without slipping. The curve traced by a point on the wheel (that starts out on the x-axis) is called a cycloid. The curve can be described parametrically by $x = \theta - \sin \theta$ and $y = 1 - \cos \theta$, where θ is the angle through which the wheel has turned. When the wheel has turned through $\pi/4$ radians, what is $\dfrac{dy}{dx}$?

7.0 For $f(x) = \arctan x$, find $f'''(x)$.

8.0 Evaluate the following limits:

$$\lim_{x \to 0} \dfrac{x - \arctan x}{x^3} \qquad \lim_{x \to \infty} \dfrac{\ln(x)}{\ln(x^2 + 1)} \qquad \lim_{x \to 0^+} (1 + x)^{\csc x}$$

8.0   Use the Mean Value theorem on the following functions, on the given intervals, if it applies:

$$f(x) = x + \sin x \text{ on } [\pi/2, \pi]$$
$$g(x) = x - x^{2/3} \text{ on } [-1, 1]$$

8.0   Suppose that f is a continuous function on $[a, b]$, differentiable on (a, b), and that $f'(x) = 0$ for all x in the interval (a, b). Show that f is a constant function on $[a, b]$.

9.0   Without using a graphing calculator, sketch graphs of these functions, showing all local extrema and inflection points:

$$g(x) = 3x^4 + 4x^3 + 1$$
$$h(x) = \ln(1 + x^2)$$

10.0   Use Newton's method to approximate a zero for the polynomial $f(x) = x^3 + 3x - 1$ in the interval $[0, 1]$. You may stop when you have a value of x for which $|f(x)| < 1/1000$.

11.0   A cone is to be made large enough to enclose a cylinder of height 5 and radius 2. What is the smallest possible volume for such a cone?

12.0   A climber on one end of a 150-foot rope has fallen down a crevasse and is slipping farther down. This accident happened because his climbing partner, on the other end of the rope, does not have a firm stance. He is on the horizontal glacier, slipping toward the crevasse at a rate of 10 ft./sec. At what rate is the distance between the two climbers changing when the first climber is 100 feet down the crevasse?

12.0   A streetlight, 20 feet in height, stands 5 feet from a sidewalk. A person, 6 feet tall, walks along the sidewalk at 4 ft./sec. At what rate is the length of the person's shadow changing when the person is 13 feet from the base of the streetlight?

14.0   A particle moves along a line with velocity function $v(t) = t^3 + t$. Find the distance traveled by the particle between times $t = 0$ and $t = 4$.

14.0   An object thrown upward in a vacuum with initial velocity V_0 will experience an acceleration of -9.8 m/s^2. Use this information to find an expression for the position of the object above its starting position after t seconds.

15.0   On the graph of $f(x) = e^{-x^2}$ shown below, let $g(s)$ denote the area under the graph of f above the x-axis, between $x = 0$ and $x = s$. Find $\lim_{s \to 0} \frac{g(s)}{s}$.

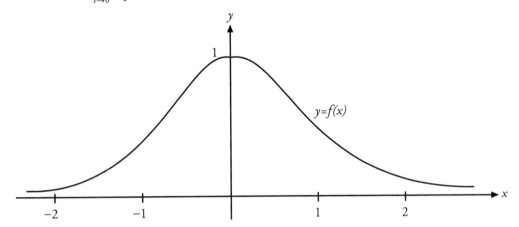

16.0   Consider the region bounded by the positive x-axis, the line $x = 1$, and the graph of $y = 1/x$. If this region is rotated about the x-axis, find the surface area of the resulting solid.

16.0   Find the length of the curve $y = (4 - x^{2/3})^{3/2}$ between $x = 1$ and $x = 8$.

16.0   The following integral represents the volume of a solid that is obtained by rotating a region in the x-y plane about one of the coordinate axes: $\pi \int_0^2 x^4 \, dx$. If this solid was obtained by rotating a region about the x-axis, then what was the region? If, on the other hand, this solid was obtained by rotating a region about the y-axis, what was the region?

16.0   The figure shown below is a cone that has been cut off at the top and then had a cone turned out of it. The radius of the top of the figure is 2 inches. The radius of the base is 4 inches. The figure is 3 inches tall. Use integration to find the volume of the figure.

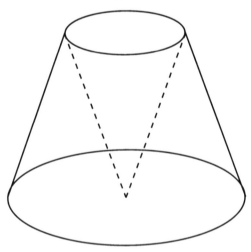

16.0 A tank of water is funnel-shaped. The shape of the funnel is such that x feet from the base of the tank, the radius of the tank is $r = \sqrt{xe^x}$ feet. If the tank is 2 feet deep and full of water, how much work is done in pumping the water out of the tank?

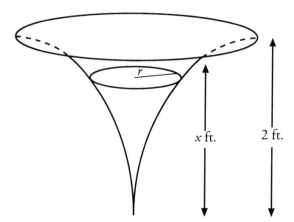

18.0 Evaluate:

$$\sin(\arctan(x))$$

$$\tan(\arcsin\sqrt{1-x^2})$$

18.0 Antidifferentiate:

$$\int \frac{1}{\sqrt{1-x^2}}\, dx$$

$$\int \frac{x}{1+x^4}\, dx$$

19.0 Evaluate:

$$\int_1^3 \frac{x^3 + 6x^2 + 13x + 8}{x^2 + 4x}\, dx$$

$$\int \frac{x^3 + 3x - 2}{x^2 + 2x + 4}\, dx$$

20.0 Evaluate:

$$\int \frac{\sin x}{\cos^3 x}\, dx$$

$$\int \sin^2 x \cos^4 x\, dx$$

21.0 Estimate $\int_0^3 e^{2x}\, dx$ using Simpson's rule with $n = 12$ subintervals and find a bound on the error.

22.0 Compute the following integrals:

$$\int_0^\pi \tan x \, dx$$

$$\int_0^\infty \frac{e^{-\sqrt{x}}}{\sqrt{x}} \, dx$$

24.0 Find the intervals of convergence for the following power series:

$$\sum_{n=2}^\infty \frac{x^n}{3^n \sqrt{n^2+1}} \qquad \sum_{n=1}^\infty \frac{(x-5)^n}{n^2-3} \qquad \sum_{n=1}^\infty \frac{2^n x^n}{(2n)!}$$

25.0 Use a Maclaurin series for the function $f(x) = \dfrac{\ln(1+x)}{x}$ to estimate

$$\int_0^{1/2} \frac{\ln(1+x)}{x} \, dx \text{ to within .01.}$$

26.0 Find the degree four Taylor polynomial for $f(x) = \sqrt{x}$, centered at $x = 1$.

26.0 Using the half-angle identity $\sin^2 x = \dfrac{1}{2}(1 - \cos(2x))$, find a Maclaurin's series for $f(x) = \sin^2 x$.

Glossary

absolute value. A number's distance from zero on the number line. The absolute value of −4 is 4; the absolute value of 4 is 4.

algorithm. An organized procedure for performing a given type of calculation or solving a given type of problem. An example is long division.

arithmetic sequence. A sequence of elements, $a_1, a_2, a_3, \ldots$, such that the difference of successive terms is a constant $a_{i+1} - a_i = k$; for example, the sequence $\{2, 5, 8, 11, 14, \ldots\}$ where the common difference is 3.

asymptotes. Straight lines that have the property of becoming and staying arbitrarily close to the curve as the distance from the origin increases to infinity. For example, the x-axis is the only asymptote to the graph of $\sin(x)/x$.

axiom. A basic assumption about a mathematical system from which theorems can be deduced. For example, the system could be the points and lines in the plane. Then an axiom would be that given any two distinct points in the plane, there is a unique line through them.

binomial. In algebra, an expression consisting of the sum or difference of two monomials (see the definition of *monomial*), such as $4a - 8b$.

binomial distribution. In probability, a binomial distribution gives the probabilities of k outcomes A (or $n-k$ outcomes B) in n independent trials for a two-outcome experiment in which the possible outcomes are denoted A and B.

binomial theorem. In mathematics, a theorem that specifies the complete expansion of a binomial raised to any positive integer power.

box-and-whisker plot. A graphical method for showing the median, quartiles, and extremes of data. A box plot shows where the data are spread out and where they are concentrated.

complex numbers. Numbers that have the form $a + bi$ where a and b are real numbers and i satisfies the equation $i^2 = -1$. Multiplication is denoted by $(a+bi)(c+di) = (ac-bd) + (ad+bc)i$, and addition is denoted by $(a+bi) + (c+di) = (a+c) + (b+d)i$.

congruent. Two shapes in the plane or in space are congruent if there is a rigid motion that identifies one with the other (see the definition of *rigid motion*).

conjecture. An educated guess.

coordinate system. A rule of correspondence by which two or more quantities locate points unambiguously and which satisfies the further property that points unambiguously determine the quantities; for example, the usual Cartesian coordinates x, y in the plane.

corollary. A direct consequence of a theorem.

cosine. Cos(θ) is the *x*-coordinate of the point on the unit circle so that the ray connecting the point with the origin makes an angle of θ with the positive *x*-axis. When θ is an angle of a right triangle, then cos(θ) is the ratio of the adjacent side with the hypotenuse.

dilation. In geometry, a transformation *D* of the plane or space is a dilation at a point *P* if it takes *P* to itself, preserves angles, multiplies distances from *P* by a positive real number *r*, and takes every ray through *P* onto itself. In case *P* is the origin for a Cartesian coordinate system in the plane, then the dilation *D* maps the point (*x*, *y*) to the point (*rx*, *ry*).

dimensional analysis. A method of manipulating unit measures algebraically to determine the proper units for a quantity computed algebraically. For example, velocity has units of the form length over time (e.g., meters per second [*m/sec*]), and acceleration has units of velocity over time; so it follows that acceleration has units (*m/sec*)/sec = $m/(sec^2)$.

expanded form. The expanded form of an algebraic expression is the *equivalent expression* without parentheses. For example, the expanded form of $(a + b)^2$ is $a^2 + 2ab + b^2$.

exponent. The power to which a number or variable is raised (the exponent may be any real number).

exponential function. A function commonly used to study growth and decay. It has the form $y = a^x$ with *a* positive.

factors. Any of two or more quantities that are multiplied together. In the expression 3.712 × 11.315, the factors are 3.712 and 11.315.

function. A correspondence in which values of one variable determine the values of another.

geometric sequence. A sequence in which there is a common ratio between successive terms. Each successive term of a geometric sequence is found by multiplying the preceding term by the common ratio. For example, in the sequence {1, 3, 9, 27, 81, . . .} the common ratio is 3.

histogram. A vertical block graph with no spaces between the blocks. It is used to represent frequency data in statistics.

heuristic argument. The term universally used in mathematics for an argument that is suggestive of the truth of a mathematical statement but which is not entirely logically correct.

hypothesis. Synonymous with *assumption*.

inequality. A relationship between two quantities indicating that one is strictly *less than* or *less than or equal* to the other.

integers. The set consisting of the positive and negative whole numbers and zero; for example, {. . . -2, -1, 0, 1, 2 . . .}.

irrational number. A number that cannot be represented as an exact ratio of two integers. For example, the square root of 2 or π.

lemma. A true statement of lesser significance than a theorem.

linear expression. An expression of the form *ax* + *b* where *x* is variable and *a* and *b* are constants; or in more variables, an expression of the form *ax* + *by* + *c*, *ax* + *by* + *cz* + *d*, etc.

linear equation. An equation containing linear expressions.

logarithm. The inverse of exponentiation; for example, $a^{(\log_a x)} = x$.

mean. In statistics, the average obtained by dividing the sum of two or more quantities by the number of these quantities.

median. In statistics, the quantity designating the middle value in a set of numbers.

mode. In statistics, the value that occurs most frequently in a given series of numbers.

monomial. In the variables x, y, z, a monomial is an expression of the form $ax^m y^n z^k$, in which m, n, and k are nonnegative integers and a is a constant (e.g., $5x^2$, $3x^2y$ or $7x^3yz^2$).

nonstandard unit. Unit of measurement expressed in terms of objects (such as paper clips, sticks of gum, shoes, etc.).

parallel. Given distinct lines in the plane that are infinite in both directions, the lines are parallel if they never meet. Two distinct lines in the coordinate plane are parallel if and only if they have the same slope.

permutation. A permutation of the set of numbers $\{1, 2, \ldots, n\}$ is a reordering of these numbers.

polar coordinates. The coordinate system for the plane based on $r\theta$, the distance from the origin and θ, and the angle between the positive x-axis and the ray from the origin to the point.

polar equation. Any relation between the polar coordinates (r, θ) of a set of points (e.g., $r = 2\cos q$ is the polar equation of a circle).

polynomial. In algebra, a sum of monomials; for example, $x^2 + 2xy + y^2$.

postulate. Synonymous with *axiom*.

prime. A natural number p greater than 1 is prime if and only if the only positive integer factors of p are 1 and p. The first seven primes are 2, 3, 5, 7, 11, 13, 17.

quadratic function. A function given by a polynomial of degree 2.

random variable. A function on a probability space.

range. In statistics, the difference between the greatest and smallest values in a data set. In mathematics, the image of a function.

ratio. A comparison expressed as a fraction. For example, there is a ratio of three boys to two girls in a class (3/2, 3:2).

rational numbers. Numbers that can be expressed as the quotient of two integers; for example, 7/3, 5/11, −5/13, 7 = 7/1.

real numbers. All rational and irrational numbers.

reflection. The reflection through a line in the plane or a plane in space is the transformation that takes each point in the plane to its mirror image with respect to the line or its mirror image with respect to the plane in space. It produces a mirror image of a geometric figure.

rigid motion. A transformation of the plane or space, which preserves distance and angles.

root extraction. Finding a number that can be used as a factor a given number of times to produce the original number; for example, the fifth root of $32 = 2$ because $2 \times 2 \times 2 \times 2 \times 2 = 32$.

rotation. A rotation in the plane through an angle θ and about a point P is a rigid motion T fixing P so that if Q is distinct from P, then the angle between the lines PQ and $PT(Q)$ is always θ. A rotation through an angle θ in space is a rigid motion T fixing the points of a line l so that it is a rotation through θ in the plane perpendicular to l through some point on l.

scalar matrix. A matrix whose diagonal elements are all equal while the nondiagonal elements are all 0. The identity matrix is an example.

scatterplot. A graph of the points representing a collection of data.

scientific notation. A shorthand way of writing very large or very small numbers. A number expressed in scientific notation is expressed as a decimal number between 1 and 10 multiplied by a power of 10 (e.g., $7000 = 7 \times 10^3$ or $0.0000019 = 1.9 \times 10^{-6}$).

sieve of Eratosthenes. A method of getting all the primes in a certain range, say from 2 to 300. Start with 2, cross out all numbers from 2 to 300 which are multiples of 2 but not equal to 2. Go to the next remaining number, which is 3. Now cross out all numbers up to 300 which are multiples of 3 but not equal to 3. Go to the next remaining number, which is 5. Cross out all remaining numbers which are multiples of 5 but not equal to 5. And so on. At each stage, the next number is always a prime. At the end of this process, when there are no more numbers below 300 to be crossed out, every remaining number is a prime. (For the case at hand, once multiples of 17 other than 17 itself have been crossed out, the process comes to an end since the product of any two primes greater than 17 must be greater than 300.)

similarity. In geometry, two shapes R and S are similar if there is a dilation D (see the definition of *dilation*) that takes S to a shape congruent to R. It follows that R and S are similar if they are congruent after one of them is expanded or shrunk.

sine. $\text{Sin}(\theta)$ is the y-coordinate of the point on the unit circle so that the ray connecting the point with the origin makes an angle of θ with the positive x-axis. When θ is an angle of a right triangle, then $sin(\theta)$ is the ratio of the opposite side with the hypotenuse.

square root. The square roots of n are all the numbers m so that $m^2 = n$. The square roots of 16 are 4 and -4. The square roots of -16 are $4\,i$ and $-4\,i$.

standard deviation. A statistic that measures the dispersion of a sample.

symmetry. A symmetry of a shape S in the plane or space is a rigid motion T that takes S onto itself $(T(S) = S)$. For example, reflection through a diagonal and a rotation through a right angle about the center are both symmetries of the square.

system of linear equations. Set of equations of the first degree (e.g., $x + y = 7$ and $x - y = 1$). A solution of a set of linear equations is a set of numbers a, b, c, ... so that when the variables are replaced by the numbers, all the equations are satisfied. For example, in the equations above, $x = 4$ and $y = 3$ is a solution.

theorem. A significant true statement in mathematics.

translation. A rigid motion of the plane or space of the form X goes to $X + V$ for a fixed vector V.

transversal. In geometry, given two or more lines in the plane a transversal is a line distinct from the original lines and intersects each of the given lines in a single point.

unit fraction. A fraction whose numerator is 1 (e.g., $1/\pi$, $1/3$, $1/x$). Every nonzero number may be written as a unit fraction since, for n not equal to 0, $n = 1/(1/n)$.

variable. A placeholder in algebraic expressions; for example, in $3x + y = 23$, x and y are variables.

vector. Quantity that has magnitude (length) and direction. It may be represented as a directed line segment.

zeros of a function. The points at which the value of a function is zero.

Works Cited

Bahrick, H. P., and L. K. Hall. 1991. "Lifetime Maintenance of High School Mathematics Content," *Journal of Experimental Psychology: General,* Vol. 120, 22–33.

Beaton, A. E., et al. 1996. *Mathematics Achievement in the Middle School Years: IEA's Third International Mathematics and Science Study (TIMSS).* Chestnut Hill, Mass.: Boston College, Center for the Study of Testing, Evaluation, and Educational Policy.

Belmont, J. M. 1989. "Cognitive Strategies and Strategic Learning: The Socio-Instructional Approach," *American Psychologist,* Vol. 44, 142–48.

Benbow, C. P., and J. C. Stanley. 1996. "Inequity in Equity: How 'Equity' Can Lead to Inequity in High-Potential Students," *Psychology, Public Policy, and Law,* Vol. 2, 249–92.

Bishop, J. H. 1989. "Is the Test Score Decline Responsible for the Productivity Growth Decline?" *American Economic Review,* Vol. 79, 178–97.

Brase, G. L.; L. Cosmides; and J. Tooby. 1998. "Individuation, Counting, and Statistical Inference: The Role of Frequency and Whole-Object Representations in Judgment Under Uncertainty," *Journal of Experimental Psychology: General,* Vol. 127, 3–21.

Note: The publication data in this section were supplied by the Curriculum Frameworks and Instructional Resources Office, California Department of Education. Questions about the references should be addressed to that office; telephone (916) 657-3023.

Briars, D., and R. S. Siegler. 1984. "A Featural Analysis of Preschoolers' Counting Knowledge," *Developmental Psychology,* Vol. 20, 607–18.

California Commission on Teacher Credentialing. 1997. *California Standards for the Teaching Profession.* Sacramento: California Commission on Teacher Credentialing.

California Education Round Table (CERT). Forthcoming. "Taking It to the Next Level: Mathematics Assessment Standards for High School Graduates." Sacramento: California Education Round Table.

California Education Round Table (CERT). 1997. *Standards in English and Mathematics for California High School Graduates.* Sacramento: California Education Round Table.

California Mathematics Task Force. 1995. *Improving Mathematics Achievement for All California Students.* Sacramento: California Department of Education.

California State University Board of Trustees. 1998. *Precollegiate Education Policy Implementation: Second Annual Report.* Agenda. March 17–18, 1998. Committee on Education Policy. Long Beach: California State University, Office of the Chancellor.

Cederberg, J. N. 1989. *A Course on Modern Geometries.* New York: Springer Verlag.

Clark, B. 1997. *Growing Up Gifted: Developing the Potential of Children at Home and at School* (Fifth edition). Needham Heights, Mass.: Prentice-Hall.

Works Cited

Cooper, G., and J. Sweller. 1987. "Effects of Schema Acquisition and Rule Automation on Mathematical Problem-Solving Transfer," *Journal of Educational Psychology*, Vol. 79, 347–62.

Cooper, H. 1989. "Synthesis of Research on Homework," *Educational Leadership*, Vol. 47, 85–91.

Delaney, P. F., et al. 1998. "The Strategy-Specific Nature of Improvement: The Power Law Applies by Strategy Within Task," *Psychological Science*, Vol. 9, 1–7.

Delcourt, M. A. B., et al. 1994. *Evaluation of the Effects of Programming Arrangements on Student Learning Outcomes*. Charlottesville: University of Virginia.

Dixon, R. C., et al. 1998. *Report to the California State Board of Education and Addendum to the Principal Report: Review of High-Quality Experimental Mathematics Research*. Eugene, Ore.: National Center to Improve the Tools of Educators.

Ericsson, K. A.; R. T. Krampe; and C. Tesch-Römer. 1993. "The Role of Deliberate Practice in the Acquisition of Expert Performance," *Psychological Review*, Vol. 100, 363–406.

Fennema, E., et al. 1981. "Increasing Women's Participation in Mathematics: An Intervention Study," *Journal for Research in Mathematics Education*, Vol. 12, 3–14.

Fuson, K. C., and Y. Kwon. 1992. "Korean Children's Understanding of Multidigit Addition and Subtraction," *Child Development*, Vol. 63, 491–506.

Geary, D. C. 1994. *Children's Mathematical Development: Research and Practical Applications*. Washington, D.C.: American Psychological Association.

Geary, D. C. 1995. "Reflections of Evolution and Culture in Children's Cognition: Implications for Mathematics Development and Mathematics Instruction," *American Psychologist*, Vol. 50, 24–27.

Geary, D. C., and K. F. Widaman. 1992. "Numerical Cognition: On the Convergence of Componential and Psychometric Models," *Intelligence*, Vol. 16, 47–80.

Geary, D. C.; C. C. Bow-Thomas; and Y. Yao. 1992. "Counting Knowledge and Skill in Cognitive Addition: A Comparison of Normal and Mathematically Disabled Children," *Journal of Experimental Child Psychology*, Vol. 54, 372–91.

Geary, D. C., et al. 1998. "A Biocultural Model of Academic Development," in *Global Prospects for Education: Development, Culture, and Schooling*. Edited by S. G. Paris and H. M. Wellman. Washington, D.C.: American Psychological Association.

Gelman, R., and E. Meck. 1983. "Preschoolers' Counting: Principles Before Skill," *Cognition*, Vol. 13, 343–59.

Goldberg, L. R. 1992. "The Development of Markers for the Big-Five Factor Structure," *Psychological Assessment*, Vol. 4, 26–42.

Greenberg, M. J. 1993. *Euclidean and Non-Euclidean Geometries* (Third edition). New York: W. H. Freeman.

Grogger, J., and E. Eide. 1995. "Changes in College Skills and the Rise in the College Wage Premium," *Journal of Human Resources*, Vol. 30, 280–310.

Holz, A. 1996. *Walking the Tightrope: Maintaining Balance for Student Achievement in Mathematics*. San Luis Obispo: California Polytechnic State University, Central Coast Mathematics Project.

Intersegmental Committee of the Academic Senates (ICAS). 1997. *Statement* on *Competencies in Mathematics Expected of Entering College Students.* Sacramento: Intersegmental Committee of the Academic Senates.

Kame'enui, E. J., and D. C. Simmons. 1998. "Beyond Effective Practice to Schools as Host Environments: Building and Sustaining a Schoolwide Intervention Model in Beginning Reading," *Oregon School Study Council Bulletin,* Vol. 41, No. 3, 3–24.

Kulik, J. A. 1992. *An Analysis of the Research on Ability Grouping: Historical and Contemporary Perspectives.* Storrs, Conn.: The National Research Center on the Gifted and Talented.

Loveless, T. 1998. "The Tracking and Ability Grouping Debate," *Fordham Report,* Vol. 2, No. 8.

Matthews, M. H. 1992. *Making Sense of Place: Children's Understanding of Large-Scale Environments.* Savage, Md.: Barnes and Noble Books.

Mayer, R. E. 1985. "Mathematical Ability," in *Human Abilities: An Information-Processing Approach.* Edited by R. J. Sternberg. San Francisco: Freeman.

Mosteller, F.; R. Light; and J. Sachs. 1996. "Sustained Inquiry in Education: Lessons from Skill Grouping and Class Size," *Harvard Educational Review,* Vol. 66, No. 4, 797–842.

National Center for Education Statistics (NCES). 1997. *Degrees and Other Awards Conferred by Institutions of Higher Education: 1994-95.* Higher Education General Information Survey (HEGIS). Washington, D.C.: U.S. Department of Education.

National Council of Teachers of Mathematics (NCTM). 1989. *Curriculum and Evaluation Standards for School Mathematics.* Reston, Va.: National Council of Teachers of Mathematics.

Nicholls, J. G. 1984. "Achievement Motivation: Conceptions of Ability, Subjective Experience, Task Choice, and Performance," *Psychological Review,* Vol. 91, 328–46.

Ohlsson, S., and E. Rees. 1991. "The Function of Conceptual Understanding in the Learning of Arithmetic Procedures," *Cognition and Instruction,* Vol. 8, 103–79.

Paglin, M., and A. M. Rufolo. 1990. "Heterogeneous Human Capital, Occupational Choice, and Male-Female Earnings Differences," *Journal of Labor Economics,* Vol. 8, 123–44.

Reese, C. M., et al. 1997. *NAEP 1996 Mathematics Report Card for the Nation and the States.* Washington, D.C.: U.S. Department of Education.

Rivera-Batiz, F. L. 1992. "Quantitative Literacy and the Likelihood of Employment Among Young Adults in the United States," *Journal of Human Resources,* Vol. 27, 313–28.

Rogers, K. B. 1991. *The Relationship of Grouping Practices to the Education of the Gifted and Talented Learner.* Storrs, Conn.: The National Research Center on the Gifted and Talented.

Saxe, G. B.; S. R. Guberman; and M. Gearhart. 1987. "Social Processes in Early Number Development," *Monographs of the Society for Research in Child Development,* Vol. 52, Serial no. 216.

Seron, X., and M. Fayol. 1994. "Number Transcoding in Children: A Functional Analysis," *British Journal of Developmental Psychology,* Vol. 12, 281–300.

Shore, B. M., et al. 1991. *Recommended Practices in Gifted Education: A Critical Analysis.* New York: Teachers College Press.

Siegler, R. S. 1995. "How Does Change Occur: A Microgenetic Study of

Number Conservation," *Cognitive Psychology,* Vol. 28, 225–73.

Siegler, R. S., and K. Crowley. 1994. "Constraints on Learning in Nonprivileged Domains," *Cognitive Psychology,* Vol. 27, 194–226.

Siegler, R. S., and E. Stern. 1998. "Conscious and Unconscious Strategy Discoveries: A Microgenetic Analysis," *Journal of Experimental Psychology: General,* Vol. 127, 377–97.

Slavin, R. E.; N. L. Karweit; and B. A. Wasik, eds. 1994. *Preventing Early School Failure: Research, Policy, and Practice.* Boston: Allyn and Bacon.

Sophian, C. 1997. "Beyond Competence: The Significance of Performance for Conceptual Development," *Cognitive Development,* Vol. 12, 281–303.

Starkey, P. 1992. "The Early Development of Numerical Reasoning," *Cognition,* Vol. 43, 93–126.

Stevenson, H. W., et al. 1990. "Contexts of Achievement: A Study of American, Chinese, and Japanese Children," *Monographs of the Society for Research in Child Development,* Vol. 55, Serial no. 221.

Stigler, J. W.; S. Y. Lee; and H. W. Stevenson. 1987. "Mathematics Classrooms in Japan, Taiwan, and the United States," *Child Development,* Vol. 58, 1272–85.

Sweller, J.; R. F. Mawer; and M. R. Ward. 1983. "Development of Expertise in Mathematical Problem Solving," *Journal of Experimental Psychology: General,* Vol. 112, 639–61.

VanLehn, K. 1990. *Mind Bugs: The Origins of Procedural Misconceptions.* Cambridge: Massachusetts Institute of Technology Press.

Wenglinsky, H. 1998. *Does It Compute?* Princeton, N.J.: Educational Testing Service, Policy Information Center.

World Math Challenge. Vol. 1. [CD-ROM]. 1995. Bellevue, Wash.: Pacific Software Publishing, Inc.

Additional References

Anderson, J. R.; L. M. Reder; and H. A. Simon. 1996. "Situated Learning and Education," *Educational Researcher,* Vol. 25, 5–11.

Ball, S. 1988. "Computers, Concrete Materials, and Teaching Fractions," *School Science and Mathematics,* Vol. 88, 470–75.

Boysen, S. T., and G. G. Berntson. 1989. "Numerical Competence in a Chimpanzee (Pan troglodytes)," *Journal of Comparative Psychology,* Vol. 103, 23–31.

Brown, R. 1973. *A First Language: The Early Stages.* Cambridge: Harvard University Press.

California Department of Education. 1996. *Teaching Reading: A Balanced, Comprehensive Approach to Teaching Reading in Prekindergarten Through Third Grade.* Sacramento: California Department of Education.

Chapins, S. 1997. *The Partners in Change Handbook: A Professional Development Curriculum in Mathematics.* Boston: Boston University.

Ferrell, B. 1986. "Evaluating the Impact of CAI on Mathematics Learning: Computer Immersion Project," *Journal of Educational Computing Research,* Vol. 2, 327–36.

Fletcher, J. D.; D. E. Hawley; and P. K. Piele. 1990. "Costs, Effects, and Utility of Microcomputer-Assisted Instruction in the Classroom. *American Educational Research Journal,* Vol. 27, 783–806.

Gaslin, W. L. 1975. "A Comparison of Achievement and Attitudes of Students Using Conventional or Calculator-Based Algorithms for Operations on Positive Rational Numbers in Ninth-Grade General Mathematics," *Journal for Research in Mathematics Education,* Vol. 6, 95–108.

Geary, D. C., et al. 1996. "Development of Arithmetical Competencies in Chinese and American Children: Influence of Age, Language, and Schooling," *Child Development,* Vol. 67, 2022–44.

Gelman, R. 1990. "First Principles Organize Attention to and Learning About Relevant Data: Number and the Animate-Inanimate Distinction as Examples." *Cognitive Science,* Vol. 14, 79–106.

Greeno, J. G.; M. S. Riley; and R. Gelman. 1984. "Conceptual Competence and Children's Counting," *Cognitive Psychology,* Vol. 16, 94–143.

Haimo, D. T. 1998. "Are the NCTM Standards Suitable for Systemic Adoption?" *Teachers College Record of Columbia University* (Special issue). Vol. 100, No. 1, 45–64.

Hatfield, L., and T. Kieren. 1972. "Computer-Assisted Problem Solving in School Mathematics," *Journal for Research in Mathematics Education,* Vol. 3, 99–112.

Hoover-Dempsey, K. V., and H. M. Sandler. 1997. "Why Do Parents Become Involved in Their Children's

Note: The publication data in this section were supplied by the Curriculum Frameworks and Instructional Resources Office, California Department of Education. Questions about the references should be addressed to that office; telephone (916) 657-3023.

Education?" *Review of Educational Research,* Vol. 67, 3–42.

Introduction to TIMSS: The Third International Mathematics and Science Study. 1997. U.S. Department of Education, Office of Educational Research and Improvement.

Johnson-Gentile, K.; D. Clements; and M. Battista. 1994. "Effects of Computer and Noncomputer Environments on Students' Conceptualizations of Geometric Motions," *Journal of Educational Computing Research,* Vol. 11, 121–40.

Kuhl, P. K., et al. 1997. "Cross-Language Analysis of Phonetic Units in Language Addressed to Infants," *Science,* Vol. 277, 684–86.

Lance, K. C. 1994. "The Impact of School Library Media Centers on Academic Achievement," *School Library Media Annual,* Vol. 12, 188–97.

LeFevre, J. A.; S. L. Greenham; and N. Waheed. 1993. "The Development of Procedural and Conceptual Knowledge in Computational Estimation," *Cognition and Instruction,* Vol. 11, 95–132.

Lewis, A. B., and R. E. Mayer. 1987. "Students' Miscomprehension of Relational Statements in Arithmetic Word Problems," *Journal of Educational Psychology,* Vol. 79, 363–71.

Mayer, R. E. 1982. "Memory for Algebra Story Problems," *Journal of Educational Psychology,* Vol. 74, 199–216.

Mayes, R. 1992. "The Effects of Using Software Tools on Mathematical Problem Solving in Secondary Schools," *School Science and Mathematics,* Vol. 92, 243–48.

McCollister, T., et al. 1986. "Effects of Computer-Assisted Instruction and Teacher-Assisted Instruction on Arithmetic Task Achievement Scores of Kindergarten Children," *Journal of Educational Research,* Vol. 80, 121–25.

Mevarech, Z.; O. Silber; and D. Fine. 1991. "Learning with Computers in Small Groups: Cognitive and Affective Outcomes," *Journal of Educational Computing Research,* Vol. 7, 233–43.

Miller, K. F., and D. R. Paredes. 1990. "Starting to Add Worse: Effects of Learning to Multiply on Children's Addition," *Cognition,* Vol. 37, 213–42.

Miller, K. F., et al. 1995. "Preschool Origins of Cross-National Differences in Mathematical Competence: The Role of Number-Naming Systems," *Psychological Science,* Vol. 6, 56–60.

Moderator's Guide to Eighth-Grade Mathematics Lessons: United States, Japan, and Germany. 1997. The Third International Mathematics and Science Study (TIMSS). Washington, D.C.: U.S. Department of Education, Office of Educational Research and Improvement.

Mullis, I. S., et al. 1997. *Benchmarking to International Achievement.* The Third International Mathematics and Science Study (TIMSS). Washington, D.C.: U.S. Department of Education, Office of Educational Research and Development.

Nastasi, B.; M. Battista; and D. Clements. 1990. "Social-Cognitive Interactions, Motivation, and Cognitive Growth in Logo Programming and CAI Problem-Solving Environments," *Journal of Educational Psychology,* Vol. 82, 150–58.

Okolo, C. M. 1992. "The Effects of Computer-Assisted Instruction: Formal and Initial Attitude on the Arithmetic Facts Proficiency and Continuing Motivation of Students with Learning Disabilities," *Exceptionality,* Vol. 3, 195–211.

Ortiz E., and S. K. MacGregor. 1991. "Effects of Logo Programming on Understanding Variables," *Journal of Educational Computing Research,* Vol. 17, 37–50.

Peterson, P. L., and E. Fennema. 1985. "Effective Teaching, Student Engagement in Classroom Activities, and Sex-Related Differences in Learning Mathematics," *American Educational Research Journal,* Vol. 22, 309–35.

Pinker, S. 1994. *The Language Instinct.* New York: William Morrow.

Pugh, K. R., et al. 1997. "Predicting Reading Performance from Neuroimaging Profiles: The Cerebral Basis of Phonological Effects in Printed-Word Identification," *Journal of Experimental Psychology: Human Perception and Performance,* Vol. 23, 299–318.

Rickard, T. C.; A. F. Healy; and L. E. Bourne, Jr. 1994. "On the Cognitive Structure of Basic Arithmetic Skills: Operation, Order, and Symbol Transfer Effects," *Journal of Experimental Psychology: Learning, Memory, and Cognition,* Vol. 20, 1139–53.

Schnur, J., and J. Lang. 1976. "Just Pushing Buttons or Learning? A Case for Mini-Calculators," *Arithmetic Teacher,* Vol. 23, 559–62.

Schoenfeld, A. 1998. "When Good Teaching Leads to Bad Results: The Disasters of Well-Taught Mathematics Courses," *Educational Psychologist,* Vol. 23, 145–66.

Shiah, S.; M. A. Mastropier; and T. E. Scruggs. 1995. "The Effects of Computer-Assisted Instruction on the Mathematical Problem Solving of Students with Learning Disabilities," *Exceptionality,* Vol. 5, 131–61.

Siegler, R. S. 1988. "Strategy Choice Procedures and the Development of Multiplication Skill," *Journal of Experimental Psychology: General,* Vol. 117, 258–75.

Siegler, R. S. 1989. "Hazards of Mental Chronometry: An Example from Children's Subtraction," *Journal of Educational Psychology,* Vol. 81, 497–506.

Siegler, R. S., and E. Jenkins. 1989. *How Children Discover New Strategies.* Hillsdale, N.J.: Erlbaum.

Standifer, C., and E. Maples. 1981. "Achievement and Attitude of Third-Grade Students Using Two Types of Calculators," *School Science and Mathematics,* Vol. 81, 17–24.

Szetela, W., and D. Super. 1987. "Calculators and Instruction in Problem Solving in Grade Seven," *Journal for Research in Mathematics Education,* Vol. 18, 215–29.

Takahira, S., et al. 1998. *Pursuing Excellence: A Study of U.S. Twelfth-Grade Mathematics and Science Achievement in International Context.* Washington, D.C.: U.S. Department of Education.

Wellman, H. M. 1990. *The Children's Theory of Mind.* Cambridge: Massachusetts Institute of Technology Press.

Wheatley, C. 1980. "Calculator Use and Problem-Solving Performance," *Journal for Research in Mathematics Education,* Vol. 11, 323–34.

Wu, H. 1996. "The Mathematician and the Mathematics Education Reform," *Notices of the American Mathematical Society* (December 1996).

Wu, H. 1999. "The 1997 Mathematics Standards War in California," in *What Is at Stake at the K–12 Standards Wars.* Edited by Sandra Stotsky. New York: Peter Lang Publishers.

Resources for Advanced Learners

The Association for Supervision and Curriculum Development
1703 N. Beauregard Street
Alexandria, VA 22311-1714
Phone: 800-933-ASCD (2723)
Fax: 703-575-5400
E-mail: member@ascd.org
Internet: www.ascd.org

The Association for Supervision and Curriculum Development (ASCD) has produced a series of professional development videotapes and training guides showing how teachers can organize instruction for advanced students.

The California Association for the Gifted
5777 W. Century Blvd., Suite 1670
Los Angeles, CA 90045
Phone: 310-215-l832

The California Association for the Gifted offers statewide advocacy and assistance.

The National Research Center on the Gifted and Talented
362 Fairfield Road, U-7
Storrs, CT 06269-2007
1-800-486-4676

The National Research Center on the Gifted and Talented offers parents and educators access to research and resources to support advanced learning in all subject areas, including mathematics.